The Monty Hall Problem

The Monty Hall Problem

The Remarkable Story of Math's Most Contentious Brainteaser

Jason Rosenhouse

UNIVERSITY PRESS
2009

OXFORD
UNIVERSITY PRESS

Oxford University Press, Inc., publishes works that further
Oxford University's objective of excellence
in research, scholarship, and education.

Oxford New York
Auckland Cape Town Dar es Salaam Hong Kong Karachi
Kuala Lumpur Madrid Melbourne Mexico City Nairobi
New Delhi Shanghai Taipei Toronto

With offices in
Argentina Austria Brazil Chile Czech Republic France Greece
Guatemala Hungary Italy Japan Poland Portugal Singapore
South Korea Switzerland Thailand Turkey Ukraine Vietnam

Published by Oxford University Press, Inc.
198 Madison Avenue, New York, NY 10016

www.oup.com

Oxford is a registered trademark of Oxford University Press

Library of Congress Cataloging-in-Publication Data
Rosenhouse, Jason.
The Monty Hall problem: the remarkable story of math's most
contentious brainteaser / Jason Rosenhouse.
p. cm.
Includes bibliographical references and index.
ISBN 978-0-19-536789-8
1. Monty Hall problem.
2. Mathematical recreations.
I. Title.
QA95.R67 2009   519.2—dc22   2009002921

9 8 7 6 5 4 3 2 1

Printed in the United States of America
on acid-free paper

To Joan Goodman and Merle Rose
for teaching me how to write
and to Dorothy Wallace
for teaching me how to do math

Contents

Preface and Acknowledgments

I began this book as a journalist. I finished it as a diarist.

In its classical form the Monty Hall problem presents you with three identical doors concealing two goats and one car. You select a door at random but do not open it. Monty now opens a door different from the one you chose, careful always to open a door that conceals a goat. We stipulate that Monty chooses his door randomly on those occasions where he has a choice (which happens when your initial choice conceals the car). He then gives you the options of sticking with your original door or switching to the remaining unopened door. You win whatever is behind your final choice. What should you do?

If you are reading this, it is likely you already know something of the history of this problem. It is the rare math problem indeed that finds its way onto the front page of the Sunday *New York Times*, but the Monty Hall problem accomplished that feat on July 21, 1991. The occasion for this attention was a heated argument that had played out over the previous eleven months in the pages of *Parade* magazine, in a Q&A column written by Marilyn vos Savant. Vos Savant presented the correct answer, that you gain a significant advantage by switching doors. There followed several rounds of heated correspondence between vos Savant and certain irate readers, many of whom had advanced degrees in mathematics, who rejected her solution.

My original idea for this book was that an entire first course in probability could be based on nothing more than variations of the Monty Hall problem. At that time my files contained a handful of papers exploring the subtleties of the classical problem and its variants, and my intent was simply to gather this material into one place. The most elementary probabilistic notions, such as defining an appropriate sample space and probability distribution or the sum rule for probabilities, arise naturally in considering the classical problem. Variations that call for Monty to behave in nonstandard ways lead immediately to conditional probabilities and Bayes' theorem. Increasing the number of doors and the number of kinds of prizes calls for a discussion of random variables and expectations. On and on it goes.

Then a funny thing happened. I discovered that the professional literature on the problem was far more vast than I had ever imagined. Even more so than mathematicians, cognitive scientists and psychologists have been making a good living trying to explain why people find this particular problem so maddeningly difficult. Philosophers have used it directly in exploring the nature of probability, and indirectly in shedding light on seemingly unrelated philosophical questions. Economists have made their own contributions to the literature, studying what the Monty Hall problem tells us about human decision making. This barely scratches the surface.

What started as a straight math book suddenly became an excursion into intellectual territory I had not previously had occasion to explore. In presenting the solutions to the classical problem and its variants, I found I could not avoid pondering the philosophical question of how best to interpret probabilistic statements. A handful of papers by physicists describing quantum mechanical versions of the problem led me to a consideration of how quantum information differed from the classical version pioneered by Claude Shannon. My collection of psychology papers led me to a vast, and often contentious, literature regarding people's facility with probabilistic reasoning.

As a result, this book is only partly about an amusing exercise in elementary probability and the sorts of mathematics needed to solve it. It is also a re-creation of the journey I took in exploring the myriad facets of the problem. It is a journey I have not yet completed, as I describe in the final chapter.

The book is organized as follows: Chapter 1 describes the history of the problem, which extends back considerably further than vos Savant's 1990 column. In Chapter 2 we solve the classical Monty Hall problem. Chapter 3 describes certain variations of the classical version that lead naturally to a discussion of Bayes' theorem and conditional probability. Chapter 4 takes up a surprisingly difficult multiple-door version. Chapter 5 is a grab bag of variations and other bits of Monty Hall esoterica that did not fit comfortably in the previous chapters. Chapter 6 surveys a few highlights from the psychological literature as we attempt once and for all to pinpoint just why people find this problem so difficult. Chapter 7 considers, and in some cases rejects, certain philosophical arguments that have the Monty Hall problem at their core. We will emerge from this discussion with a deeper appreciation of the problem's fine points. There is then a short concluding chapter, after which we will call it a day.

The level of mathematics varies considerably from chapter to chapter. Chapters 1, 6, and 7, as well as the latter portion of Chapter 5, contain very little mathematics. Chapter 2 requires only the most basic elements of probability theory, all of which are presented in the chapter itself. Things grow somewhat more difficult in Chapter 3, where conditional probability and Bayes' theorem make their first appearance. Though I explain all of the necessary mathematics in the text, I suspect that people with no prior familiarity with these ideas will have difficulty with portions of the chapter. Finally, Chapter 4 sees the most complex mathematics of the book. The equations get rather dense, I am afraid. We also make use of a proof by induction, elementary aspects of recurrence relations, and certain properties of the number e. Even here, however, I hope I have provided enough commentary to illuminate the gist of the argument, even for those who choose not to parse the equations. Throughout, an undergraduate mathematics major should not have much difficulty understanding anything I have presented.

Acknowledgments

A great many people provided invaluable help to me during the preparation of this book.

The mathematical arguments presented in Chapter 4 are not original to me. The first solution to the progressive Monty Hall problem presented in Chapter 4 was provided by Andrew Schepler, in response to an essay on the subject I posted at my blog. The second solution, based on recurrence relations, is based heavily on ideas developed by my James Madison University colleague Stephen Lucas. Stephen also carried out the Monte Carlo simulation referred to at the end of section 7.5. The tree diagrams in sections 2.7 and 3.9 were produced by Carl Droms.

Martin Gardner provided a lot of helpful information when I was researching the history of the problem. Like a lot of mathematicians my age (mid-thirties), some of my earliest and most pleasant mathematical experiences came from his writing. In my case it was his book *Puzzles from Other Worlds* that really sparked my interest. It has been an honor to correspond with him.

Many people read and commented on all or part of the first draft of the manuscript. There is no question that this feedback has resulted in a much better book, and I would like to thank those people here: Alan Baker, Peter Baumann, Arthur Benjamin, Paul Moser, Jonathan Lubin, Stephen Lucas, Jason Martin, Phillip Martin, David Neel, Michael Orrison, Allen Schwenk, Jeffrey Shallit, Jan Sprenger, Laura Taalman, and Anthony Tongen. Of course, it goes without saying that the fault for any remaining errors or stylistic infelicities belongs solely with me.

I would also like to thank Michael Penn of Oxford University Press for his enthusiastic support for this project after I nervously pitched the idea to him during a chance meeting at the annual Joint Mathematics Conference in 2007. When he left OUP during the writing of the book, Ned Sears and Phyllis Cohen took over flawlessly and saw things through to the end.

The Monty Hall Problem

1

Ancestral Monty

1.1. A Mathematician's Life

Like all professional mathematicians, I take it for granted that most people will be bored and intimidated by what I do for a living. Math, after all, is the sole academic subject about which people brag of their ineptitude. "Oh," says the typical well-meaning fellow making idle chitchat at some social gathering, "I was never any good at math." Then he smiles sheepishly, secure in the knowledge that his innumeracy in some way reflects well on him. I have my world-weary stock answers to such statements. Usually I say, "Well, maybe you just never had the right teacher." That defuses the situation nicely.

It is the rare person who fails to see humor in assigning to me the task of dividing up a check at a restaurant. You know, because I'm a mathematician. Like the elementary arithmetic used in check division is some sort of novelty act they train you for in graduate school. I used to reply with "Dividing up a check is applied math. I'm a pure mathematician," but this elicits puzzled looks from those who thought mathematics was divided primarily into the courses they were forced to take in order to graduate versus the ones they could mercifully ignore. I find "Better have someone else do it. I'm not good with numbers" works pretty well.

I no longer grow vexed by those who ask, with perfect sincerity, how folks continue to do mathematical research when surely everything has been figured out by now. My patience is boundless for those who assure me that their grade-school nephew is quite the little math prodigy. When a student, after absorbing a scintillating presentation of, say, the mean-value theorem, asks me with disgust what it is good for, it does not even occur to me to grow annoyed. Instead I launch into a discourse about all of the practical benefits that accrue from an understanding of calculus. ("You know how when you flip a switch the lights come on? Ever wonder why that is? It's because some really smart scientists like James Clerk Maxwell knew lots of calculus and figured out how to apply it to the problem of taming electricity. Kind of puts your whining into perspective, wouldn't you say?") And upon learning that a mainstream movie has a mathematician character, I feel cheated if that character and his profession are presented with any element of realism.

(Speaking of which, do you remember that 1966 Alfred Hitchcock movie *Torn Curtain*, the one where physicist Paul Newman goes to Leipzig in an attempt to elicit certain German military secrets? Remember the scene where Newman starts writing equations on a chalkboard, only to have an impatient East German scientist, disgusted by the primitive state of American physics, cut him off and finish the equations for him? Well, we don't do that. We don't finish each other's equations. And that scene in *Good Will Hunting* where emotionally troubled math genius Matt Damon and Fields Medalist Stellan Skarsgård high-five each other after successfully performing some feat of elementary algebra? We don't do that either. And don't even get me started on Jeff Goldblum in *Jurassic Park* or Russell Crowe in *A Beautiful Mind*.)

I tolerate these things because for all the petty annoyances resulting from society's impatience with math and science, being a mathematician has some considerable compensating advantages. My professional life is roughly equal parts doing mathematics and telling occasionally interested undergraduates about mathematics, which if you like math (and I *really* like math) is a fine professional life indeed. There is the pleasure of seeing the raised eyebrows on people's faces when it dawns on them that since I am a mathematician I must have a PhD in the subject, which in turn means that I am very, very smart. And then there is the deference I am given when the conversation turns to topics of math and science (which it often does when I am in the room). That's rather pleasant. Social conventions being what they are, it is quite rare that my opinion on number-related questions is challenged.

Unless, that is, we are discussing the Monty Hall problem.

In this little teaser we are asked to play the role of a game show contestant confronted with three identical doors. Behind one is a car; behind the other two are goats. The host of the show, referred to as Monty Hall, asks us to pick one of the doors. We choose a door but do not open it. Monty now opens a door different from our initial choice, careful always to open a door he knows to conceal a goat. We stipulate that if Monty has a choice of doors to open, then he chooses randomly from among his options. Monty now gives us the options of either sticking with our original choice or switching to the one other unopened door. After making our decision, we win whatever is behind our door. Assuming that our goal is to maximize our chances of winning the car, what decision should we make?

So simple a scenario! And apparently with a correspondingly simple resolution. After Monty eliminates one of the doors, you see, we are left with a mere two options. As far as we are concerned, these two options are equally likely to conceal the car. It follows that there is no advantage to be gained one way or the other from sticking or switching, and consequently it makes no difference what decision we make. How sneaky to throw in that irrelevant nonsense about Monty choosing randomly when he has a choice! Surely you did not expect your little mathematical mind games to work on one so perspicacious as myself!

Or so it usually goes.

The minutiae of working out precisely why that intuitive and plausible argument is nonetheless incorrect will occupy us in the next chapter. For now I will simply note that it takes a person of rare sangfroid to respond with patience and humility on being told that the correct answer is to switch doors. You can share with a college class the glories of the human intellect, the most beautiful theorems and sublime constructs ever to spring forth from three pounds of matter in a human skull, and they will dutifully jot it all down in their notes without a trace of passion. But tell them that you double your chances of winning by switching doors, and suddenly the swords are drawn and the temperature drops ten degrees.

That PhD with which they were formerly so impressed? Forgotten! The possibility that they have overlooked some subtle point in their knee-jerk reaction to the problem? Never crosses their mind! They will explain with as much patience as they can muster, as though they were now the teacher and I the student, that *it makes no difference* what door you chose originally, or how Monty chose his door to open. It matters only that *just two doors remain* after Monty does his thing. Those doors, and this is the really important part, *have an equal probability of being correct!* And when I stubbornly refuse to accept their cogent logic, when I try to explain instead that there is indeed relevance to the fact that Monty follows a particular procedure in choosing his door, the chief emotion quickly shifts from anger to pity.

My remarks thus far may have given the impression that I find this reaction annoying. Quite the contrary, I assure you. My true emotion in these situations is delighted surprise. I have presented the problem to numerous college classes and in countless other social gatherings. No matter how many times I do so, I remain amazed by the ability of a mere math problem to awaken such passion and interest. The reason, I believe, is that the Monty Hall problem does not *look* like a math problem, at least not to people who think tedious symbol manipulation is what mathematics is really all about. The problem features no mathematical symbols, no excessively abstract terminology or ideas. Indeed, the problem can be explained to a middle-school student. The scenario it describes is one in which we can all imagine ourselves. And in such a situation, why should the egghead have any advantage over the normal folks?

1.2. Probability Is Hard

The Monty Hall problem is a fine illustration of the difficulties most people encounter in trying to reason about uncertainty. Probabilistic reasoning is just not something that comes naturally. For myself, I remember the precise moment I came to realize that probability is hard. I was in high school, and my father proposed to me the following brainteaser (as we shall see throughout this book, my father often gave me puzzling problems to think about during my formative years): Imagine two ordinary, well-shuffled decks of cards on the table in front of you. Turn over the top card on each deck. What is the probability that at least one of those cards is the ace of spades?

Surely, I thought, we should reason that since a deck has fifty-two cards, only one of which is the ace of spades, the probability of getting the ace on the first deck is $\frac{1}{52}$. The probability of getting the ace of spades on the second deck is likewise $\frac{1}{52}$. Since the two decks are independent of each other, the probability that at least one of those two cards was the ace of spades should be obtained by adding the two fractions together, thereby obtaining $\frac{1}{52} + \frac{1}{52} = \frac{1}{26}$. I gave my answer.

And that is when my father gave me the bad news. My answer was incorrect since I had not adequately considered the possibility of getting the ace of spades on both decks simultaneously. Why, though, should that be relevant? If we get the two aces simultaneously, that is like a super victory! Getting them both may be overkill since only one ace was required, but it is not at all clear why that ought to alter my estimation of the probability.

Look a bit more closely, however, and you see that something is amiss. Imagine that instead of using fifty-two-card decks, we use decks containing a mere two cards. For concreteness let us assume we have removed the ace of spades and the king of hearts from each of two decks, and from now on we will work solely with these four cards. Now we have two small decks of two cards each. Let us repeat the experiment. There is now a probability of $\frac{1}{2}$ that the top card on the first deck is the ace of spades (and a corresponding probability of $\frac{1}{2}$ that the top card is the king of hearts). There is likewise a probability of $\frac{1}{2}$ that the top card of the second deck is the ace. Following our previous argument, we would now claim that the probability of obtaining the ace of spades on at least one of the two decks is given by $\frac{1}{2} + \frac{1}{2} = 1$, which would imply that we are certain to get an ace on at least one of the decks. This is plainly false, since there is a possibility of flipping up two kings.

Let us try a more rigorous argument. Notice that there are $52 \times 52 = 2{,}704$ different pairs of cards that can be formed by taking one card out of the first deck and one card out of the second deck (note that we are thinking of ordered pairs here, so that removing the two of spades from the first deck and the three of hearts from the second deck should be regarded as different from choosing the three of hearts from the first deck and the two of spades from the second). Now, there are 52 pairs where the ace of spades is the card chosen from the first deck (that is, the ace of spades from the first deck can be paired with any of the 52 cards from the second deck). There are likewise 52 pairs in which the card drawn from the second deck is the ace of spades. This makes a total of 104 out of 2,704 pairs in which the ace of spades appears. Sadly, there is one pair that has been counted twice. Specifically, the pair in which both cards are the ace of spades has been double-counted. Consequently, there are only 103 pairs of cards in which at least one of the cards is the ace of spades. This gives us a probability of $\frac{103}{2,704}$, and that is our final answer.

This was a humbling experience for me. My father's scenario was superficially very simple—just two decks of cards and a straightforward procedure. Yet a full understanding of what was going on required some careful analysis. Even after seeing the cold equations, the counterintuitive nature of the solution remains. That made quite an impression.

But for all of that, the Monty Hall problem looks at the two-deck scenario and just laughs its head off. If the two-deck problem struck you as frustratingly subtle, then there is a real danger that the Monty Hall problem will drive you insane. As much as I want people to read my book, I must advise you to consider turning back now.

1.3. The Perils of Intuition

It is customary for books about probability to try to persuade otherwise intelligent people that they are lousy when it comes to reasoning about uncertainty. There are numerous well-known examples to choose from in that regard. Since I see no reason to break from so fine a custom, I will present a few of my favorites below.

In presenting the Monty Hall problem to students I have found the common reactions to follow the well-known five stages of grief. There is denial ("There is no advantage to switching"), anger ("How dare you suggest there is an advantage to switching!"), bargaining ("It's really all a matter of perspective, so maybe we're both right"), depression ("Whatever. I'm probably wrong"), and acceptance ("Is this going to be on the test?"). I figure I can speed that process along by showing you at the outset that your intuitions about probability are sometimes mistaken. Below are three of my favorites. (I realize, of course, that the first two examples in particular are so famous that it is possible you have heard them before. If that is the case, then I apologize. But there are reasons they are classics!)

1.3.1. The Birthday Problem

Let us begin with an old chestnut known as the birthday problem. How many people do you need to assemble before the probability is greater than $\frac{1}{2}$ that some two of them have the same birthday? While you are pondering that, let me mention that I am not asking trick questions. You can safely assume there is no pair of identical twins among the people under discussion, you do not have to worry about leap years, every day is as likely as any other to be someone's birthday, and our birthdays consist of a month and a day with no year attached.

In the interest of putting a bit more space between our statement of the problem and its eventual solution, let me mention that the assumption that every day is as likely as any other to be someone's birthday is known to be unrealistic. For example, large numbers of children are conceived in the period between Christmas and New Year's, which leads to an unusually large number of children being born in September. Furthermore, many children are birthed via Caesarian section or induced labor, and these procedures are not generally scheduled for the weekends. This leads to an unusually large number of children being born on Mondays or Tuesdays (which is highly relevant for the teacher presenting this problem to a roomful of schoolchildren, since in that case most of the people in the room share a birth year).

Back to the problem. Intuition tells us the answer should be fairly large. My birthday, for example, is April 26. If I start asking random people on the street, I would expect to meet quite a few before finding another April 26 birthday. The number 183 suggests itself as an answer, since then I might informally expect to find half the dates in a calendar year represented.

The question, however, said nothing about matching the birthday of any specific person. It asked only that we have some two people provide a match. To see the distinction, suppose we have four people: Alice, Benjamin, Carol, and Dennis. Then there are only three other people who might have the same birthday as Alice. But there are six chances to find a matching pair: (Alice, Benjamin), (Alice, Carol), (Alice, Dennis), (Benjamin, Carol), (Benjamin, Dennis), (Carol, Dennis). Six chances for a matching pair versus only three for matching a specific person. Quite a difference!

To solve the problem, let us assume there are only two people, Alice and Benjamin. Then the probability is $\frac{364}{365}$ that they will have different birthdays. This follows from the observation that there are 364 days in the year that are not Alice's birthday, and each is as likely as any other to be Benjamin's birthday. If we add a third person, Carol, to the mix, there is a probability of $\frac{363}{365}$ that she will have a birthday different from Alice and Benjamin (because there are 363 days in the year that are the birthday of neither Alice nor Benjamin). And if a fourth person now enters the room, the probability that his birthday is different from everyone else in the room will be $\frac{362}{365}$. The probability P that in a roomful of n people no two of them will have the same birthday is obtained by multiplying these numbers together, giving us the formula

$$P = \left(\frac{364}{365}\right)\left(\frac{363}{365}\right)\left(\frac{362}{365}\right)\cdots\left(\frac{365 - n + 1}{365}\right).$$

This, recall, is the probability that we do not have two people with the same birthday. To answer the original question we must find the smallest value of n for which $1 - P \geq \frac{1}{2}$. It turns out that $n = 23$ does the trick.

So just 23 people are needed to have a probability greater than $\frac{1}{2}$ of having two with the same birthday. With 23 people there are 253 pairs, which means 253 chances of getting a match. Remarkable. If you are curious, it turns out that with 88 people, the probability is greater than $\frac{1}{2}$ of having three people with the same birthday, while 187 people gives a probability greater than $\frac{1}{2}$ of four people having the same birthday. These facts are rather difficult to prove, however, and I will refer you to [59] for the full details.

1.3.2. False Positives in Medical Testing

Imagine a disease which afflicts roughly one out of every one thousand members of the population. There is a test for the disease, and this test is 95% accurate. It never gives a false negative; if it says you do not have the disease, then you do not have it. But 5% of the people will test positive despite not having the disease. Let us suppose you have tested positive. What is the probability that you actually have the disease?

If you are inclined to answer that the probability is very high, then your reasoning was probably that 95% is pretty close to certain and that is the end of the story. Overlooked in this argument, however, is the significance of the disease being very rare in the population. Only 5% of the positives returned by the test are false, but the disease only afflicts $\frac{1}{10}$ of 1% of the population. Your chance of having the disease is so small to begin with that it should take something truly extraordinary to send you into a panic.

To see this, suppose that 100,000 people take the test. Of these, we expect that 100 have the disease and 99,900 do not have it (100 being .1% of 100,000). Since the test does not return false negatives, we know these 100 people will receive a positive test result. In the population of healthy people, we expect that 5%, or 4,995, will receive a false positive. That makes a total of 100 true positives out of 5,095 positive test results. Dividing 100 by 5,095 gives us something just under 2%. Testing positive barely rates an eyebrow raise, much less a fit of panic.

Indeed, even if the test were 99.9% accurate, it would still be just fifty-fifty that you have the disease (for in that case you would have .1% of 99,900 people, or 99.9, receiving false positives; that makes 200 positive tests, of which 100 are accurate and 100 are not).

This result is so surprising that we are in danger of thinking that the positive test result is essentially worthless as evidence of having the disease. This, however, would be the wrong conclusion. The positive test result changed our assessment of the probability from .1% to 2%, a twenty-fold increase. It is simply that the probability of having the disease was so small to begin with that even this increase is insufficient to make it seem likely. The tendency of people to ignore such considerations is referred to by psychologists and cognitive scientists as the "base-rate fallacy" (though I should mention there is some controversy over the extent to which people fall prey to this fallacy; see [9]).

The source of the confusion lies in misapprehending the reference class to which the 5% applies. The 95% accuracy rate means that the huge majority of people who take the test get an accurate result. Most people will test negative and will, in fact, be negative. These, however, are not the folks of interest to you upon testing positive. Instead you ought to concern yourself solely with the people who tested positive, and most of *those* people received a false result. The disease is so rare that a positive test result is far more likely to indicate an error than it is to indicate actual sickness.

Whole books get written exposing these sorts of pitfalls, and I recommend *Struck by Lightning: The Curious World of Probabilities*, by Jeffrey Rosenthal [79], as an especially good representative of the genre.

1.3.3. Thirty Percent Chance of Rain?

Sometimes the difficulty lies in the ambiguity of a probabilistic statement, as opposed to some error in reasoning. Take, for example, the weatherman's assertion that there is a 30% chance of rain tomorrow. What, exactly, does this mean?

In [32], a group of researchers polled residents of five different cities on precisely that question. The cities were New York, Amsterdam, Berlin, Milan, and Athens. New York, you see, had introduced probabilistic weather forecasts in 1965. Amsterdam and Berlin did so in 1975 and the late eighties, respectively. In Milan they are used only on the Internet, and in Athens they are not used at all. This provided some diversity in the level of exposure of the people of those cities to these sorts of forecasts.

Respondents were asked to assess which of the following three choices was the most likely, and which the least likely, to be the correct interpretation of the forecast:

1. It will rain tomorrow in 30% of the region.

2. It will rain tomorrow for 30% of the time.

3. It will rain on 30% of the days like tomorrow.

The correct answer is number three, though, in fairness, the wording is not quite right. A 30% chance of rain means roughly that in 30% of the cases when the weather conditions are like today, there was some significant amount of rain the following day. It was only in New York that a majority of respondents answered correctly. Option one was selected as most likely in each of the European cities.

The article goes on to discuss various cultural reasons for their findings, as well as some suggestions for how weather-reporting bureaus can help clear up the confusion. For us, however, the take-away message is that a probabilistic statement is ambiguous unless a clearly defined reference class is stipulated. There is nothing inherently foolish in any of the three interpretations above. A failure to pay attention to details, however, is a major source of error in probabilistic reasoning.

Along the same lines, the article [32] also contains the following amusing anecdote:

> A psychiatrist who prescribed Prozac to depressed patients used to inform them that they had a 30%–50% chance of developing a sexual problem such as impotence or loss of sexual interest. On hearing this, many patients became concerned and anxious. Eventually, the psychiatrist changed his method of communicating risks, telling patients that out of every ten people to whom he prescribes Prozac, three to five experience sexual problems. This way of communicating the risk of side effects seemed to put patients more at ease, and it occurred to the psychiatrist that he had never checked how his patients understood what a "30%–50% chance of developing a sexual problem" means. It turned out that many had thought that something would go awry in 30%–50% of their sexual encounters. The psychiatrist's original approach to risk communication left the reference class unclear: Does the percentage refer to a class of people (patients who take Prozac), to a class of events (a given person's sexual encounters), or to some other class? Whereas the psychiatrist's reference class was the total number of his patients who take Prozac, his patients' reference class was their own sexual encounters. When risks are solely communicated in terms of single-event probabilities, people have little choice but to fill in a class spontaneously based on their own perspective on the situation. Thus, single-event probability statements invite a type of misunderstanding that is likely to go unnoticed.

And if *that* does not impress upon you the importance of thinking clearly about probability, then I do not know what will!

1.4. The Legacy of Pascal and Fermat

Who is responsible for foisting on decent folks all of this subtlety and counterintuition? There is a story in that, a small part of which will now be related.

The branch of mathematics devoted to analyzing problems of chance is known as probability theory. If it strikes you as odd that mathematics, a tool devised for explicating the regularities of nature, has any light to shed on unpredictable events, then rest assured you are in good company. Probability is a relative latecomer on the mathematical scene, and as recently as 1866 the

British mathematician and philosopher John Venn (of Venn diagram fame) could write [94], without fear of being contradicted,

> To many persons the mention of Probability suggests little else than the notion of a set of rules, very ingenious and profound rules no doubt, with which mathematicians amuse themselves by setting and solving puzzles.

Bertrand Russell expressed the paradox at the heart of probability by asking rhetorically, "How dare we speak of the laws of chance? Is not chance the antithesis of all law?" A resolution to this paradox begins with the observation that while individual events are frequently unpredictable, long series of the same kind of event can be a different matter entirely. The result of a single coin toss cannot be predicted. But we can say with confidence that in a million tosses of a fair coin, the ratio of heads to tails will be very close to one.

That long-run frequencies can be stable where individual occurrences are not is a fact obvious to any gambler, and it is perhaps for this reason that probability emerged from a consideration of certain games of chance. Indeed, we now risk facing a different paradox. Since evidence of gambling goes back almost as far as human civilization itself, we might wonder why the mathematics of probability took so long to appear.

There were halting steps in that direction throughout the Middle Ages. This appears most notably in the work of Cardano and Galileo, both of whom noted that, in a variety of situations, there is insight to be gained from enumerating all of the possible outcomes from some particular chance scenario and assigning an equal probability to each. That said, it is fair to observe that probability in its modern form was born from a correspondence between the seventeenth-century French mathematicians Blaise Pascal and Pierre de Fermat. The correspondence was the result of certain questions posed to Pascal by the Chevalier de Méré, a nobleman and gambler.

Especially significant was the "problem of points." The general question is this: We have two players involved in a game of chance. The object of the game is to accumulate points. Each point is awarded in such a way that the players have equal chances of winning each point. The winner is the first player to reach a set number of points, and the prize is a pool of money to which both players have contributed equally. Suppose the game is interrupted prior to its completion. Given the score at the moment of the interruption, how ought the prize money be apportioned?

With the appearance of the word "ought" in the statement of the problem, we realize that this is not a question solely of mathematics. Some notion of fair play must be introduced to justify any proposed division. We will not delve into this aspect of things, preferring instead to use the principle that we know a fair division when we see one.

As a simple example, imagine the players are Alistair and Bernard. Points are awarded by the toss of a coin, with heads going to Alistair and tails going to Bernard. Let us say the winner is the first one to ten points, and the score is currently eight for Alistair and seven for Bernard. This is roughly the scenario pondered by Pascal and Fermat.

Fermat got the ball rolling by noting that the game will surely end after no more than four further tosses of the coin. This corresponds to sixteen scenarios, all of them equally likely. Since it is readily seen, by listing all of the possibilities, that eleven of these scenarios lead to victories for Alistair while a mere five lead to victories for Bernard, the prize money should be divided in the ratio of 11:5, with the larger share going to Alistair.

Not much to criticize there, but Pascal one-upped him by noting that enumerating all of the possibilities becomes tedious in a hurry, and can become effectively impossible for large numbers. What is needed is a general method for counting the number of scenarios in which each of the players win. He pictured a situation in which Alistair needed n points to win, while Bernard needed m points. In that case the game would end in no more than $n + m - 1$ plays, for a total of 2^{n+m-1} possible scenarios. You can imagine listing all of these possible scenarios, recording A for an Alistair point and B for a point to Bernard. The result will be 2^{n+m-1} strings of A's and B's. Any string in which $n, n + 1, n + 2, \ldots, n + m - 1$ plays come up in favor of Alistair will correspond to an Alistair victory. In modern terminology we would say that we seek the sum of the following binomial coefficients:

$$\binom{n + m - 1}{n}, \quad \binom{n + m - 1}{n + 1}, \quad \binom{n + m - 1}{n + 2}, \quad \ldots, \quad \binom{n + m - 1}{n + m - 1}.$$

Pascal's name was already associated with such objects, as he had previously written extensively about them. In particular, he devoted considerable attention to what is now known as Pascal's triangle (though the existence of the triangle was known well before Pascal arrived on the scene). The values of the binomial coefficients can be arranged in a triangle, the first five rows of which are as follows:

$$
\begin{array}{ccccccccc}
 & & & & 1 & & & & \\
 & & & 1 & & 1 & & & \\
 & & 1 & & 2 & & 1 & & \\
 & 1 & & 3 & & 3 & & 1 & \\
1 & & 4 & & 6 & & 4 & & 1 \\
\end{array}
$$

Armed with this triangle, we find row $n + m$, and sum up the first m numbers we find there. This will give us the number of scenarios in which Alistair wins.

For example, in our original game we saw that Alistair needed two more points to win, while Bernard needed three. Consequently, we have $n = 2$ and $m = 3$. It follows that the game will be over in no more than $2 + 3 - 1 = 4$ subsequent plays. To determine the number of scenarios in which Alistair wins, we go to row $n + m = 5$ of the triangle. We find the numbers 1, 4, 6, 4, 1. The sum of the first m of these entries is $1 + 4 + 6 = 11$, which is precisely what Fermat found with his less elegant approach.

Nowadays these are the sorts of considerations that appear very early in any standard undergraduate course in probability or statistics. At the time, however, this was impressive stuff. It represents one of the first serious attempts to develop a calculus for probabilities, and the level of algebraic sophistication achieved by Pascal and Fermat went far beyond anything that had been seen previously. These basic principles prove to be adequate for solving the classical Monty Hall problem.

Were this intended to be a history of probability generally, I would here take note of the many people either contemporary with or prior to Pascal and Fermat who contributed to the then-nascent theory of probability. If you are interested in such a history. I recommend either Ian Hacking's *The Emergence of Probability* [40] or F. N. David's *Games, Gods, and Gambling* [17]. (David's book is especially interesting for her opinion, in defiance of the consensus view, that Pascal's contributions have been overrated, and that Fermat deserves more credit than he gets.) But since my actual intention is to find someone to blame for the endless stream of counterintuitive probabilistic brainteasers with which generations of undergraduate math majors have been tormented, I will follow convention and blame the correspondence between Pascal and Fermat.

1.5. What Bayes Wrought

Thus far we have been concerned primarily with the problem of inferring the effects of known causes. For example, given what we know about coins, what are we likely to observe if one is tossed multiple times? Likewise for the problem of points. We are given much regarding the structure of the game, and seek some reasonable conclusion as to how things are likely to proceed.

This sort of thinking, however, can be turned around. Sometimes we have known effects and wish to work backward to what caused them. Typically there are many plausible causes for given mysterious effects. We seek some statement about which of them is most likely to be correct. This is referred to as the problem of inverse probability, and it was pioneered by an eighteenth-century British mathematician and Presbyterian minister named Thomas Bayes. His discussion of the problem appears in an essay published posthumously in 1764 entitled "An Essay Towards Solving a Problem in the Doctrine of Chances."

Upon resolving to include in this book some material on the history of probability, I thought it might be fun to read Bayes' essay. I was mistaken. Even with several modern commentaries to guide me, I found it largely impenetrable. Bayes' writing is frequently muddled and confusing, and you will search his essay in vain for anything that looks like what we now call Bayes' theorem. (A special case of the theorem appears in Bayes' essay. The modern form of Bayes' theorem received its first careful formulation in the work of Laplace.) Add to this the inevitable difficulties that arise in trying to read technical papers written long ago, at a time when much modern terminology and notation had yet to arrive on the scene, and I would say you are better off with a modern textbook.

Bayes occupies a curious position in the history of mathematics. His name is today attached to a major school of philosophical thought on the nature of probability (more on that in Chapter 3). "Bayesianism" also refers to an influential view of proper statistical reasoning. Nevertheless, histories of probability cannot seem to dismiss Bayes quickly enough. David's history of the early days of probability [17] contains not a single reference to him. Hacking mentions him only briefly in [40]. Isaac Todhunter's magisterial and still authoritative 1865 book *History of the Theory of Probability from the Time of Pascal to That of Laplace* [91] devotes a chapter to Bayes, but at six pages long it is the shortest in the book. It also quite critical of Bayes' work.

The mathematical details of Bayes' theorem will occupy us in Chapter 3. For now let us consider the more general question of how to update a prior probability assessment in the face of new evidence. In the Monty Hall problem, for example, let us assume that we initially choose door number one. Since the doors are assumed to be identical, we assign a probability of $\frac{1}{3}$ to this door. We now see Monty open one of the other two doors to reveal a goat. The question is whether our $\frac{1}{3}$ probability assignment ought to change in the light of this new information.

It will be useful, in pondering such situations, to change our perspective regarding the nature of probability. To this point we have behaved as though the point of probability was to discover the properties of certain real-world objects. Assigning a probability of $\frac{1}{2}$ to the result of a coin toss was viewed as a statement about coins, for example. More specifically, it was a description of something coins tend to do when they are flipped a large number of times. This, however, is not the only way of viewing things. We might also think of probability assignments as representing our degree of belief in a given proposition. In this view, the assignment of $\frac{1}{2}$ to each possible result of a coin toss means that we have no basis for believing that one outcome is more likely than another. It is a statement about our beliefs, as opposed to a statement about coins.

We now ask for the variables affecting how we update our degree of belief in a proposition in the face of new evidence. One consideration. I suggest, is obvious. Our updated assignment will depend in part on our prior assignment. Scientists have a saying that extraordinary claims require extraordinary evidence. This captures the insight that if we initially view a proposition as exceedingly unlikely, it will take impressive evidence indeed to make it suddenly seem likely.

The next consideration is less obvious. If A is the proposition whose probability we are trying to assess, and B is the new evidence, we want to know how tightly correlated B occurrences are with A occurrences. That is, we need to know how likely it is that B will occur given that A is true. If we assess this probability as very high, then the occurrence of B will increase our confidence in the truth of A. Perhaps we are on a jury in a criminal trial. We learn that a hair found at the scene of the crime provides a DNA match with the defendant. Since the probability that such evidence will be found given that the suspect is guilty is quite high, the DNA match would tend to increase our confidence in the guilt of the defendant. But now suppose we learn that the suspect has an unbreakable alibi for the time of the crime. Since the probability that the defendant would have such an alibi given that he is guilty is quite low, this revelation would decrease our confidence in the guilt of the defendant.

This, however, cannot be the end of the story. It is possible that B is the sort of thing that happens frequently regardless of whether or not A is true. In such a situation we would assess that the probability of B given that A is true is high not because of any particular connection between A and B, but simply because B is something that is very likely to happen regardless. The finding that B is very likely to happen diminishes its relevance as evidence for A. What matters, then, is not just how likely it is that B will occur given that A is true. Rather, we seek the ratio of this probability to the probability that B will occur barring any assumption about A.

In other words, B should be viewed as strong evidence in support of A if B is something that is likely to occur if A is true, but unlikely to occur if A is false. Let us suppose our defendant has no plausible reason for being present at the scene of the crime. Then we might say that finding his DNA at the scene is likely to happen only if he is guilty, and the DNA match is strong evidence for the prosecution. But if it turns out that the crime was committed in a place the defendant often frequents for entirely innocent reasons, then the DNA match is likely to occur independent of any assumption about his guilt. In this case, the DNA match constitutes weak evidence indeed.

Bayes' theorem takes these vague intuitions and turns them into a precise formula for updating prior probability assessments. It will be our constant companion through most of this book.

1.6. The Bertrand Box Paradox

The Monty Hall problem in its modern form goes back to 1975, and I assure you we will arrive at *that* little matter soon enough. During the prior sixteen years it was traveling incognito as the three-prisoners problem. That will be the subject of the next section.

Annoying brainteasers in conditional probability, however, have a far longer history, and there is one little bagatelle of sufficient importance to rate a mention in this chapter. It is nowadays referred to as the Bertrand box paradox, in honor of French mathematician Joseph Bertrand. It appeared in his 1889 book *Calcul des Probabilités* (Calculus of Probabilities) as follows (see [69] for a useful discussion of Bertrand's thinking):

Three boxes are identical in external appearance. The first box contains two gold coins, the second contains two silver coins, and the third contains a coin of each kind, one gold and one silver. A box is chosen at random. What is the probability that it contains the unlike coins?

If your first instinct was that the problem is trivial, but then you worried that if it were *really* trivial I would not have included it, then rest assured that this is one of the few places in the book where you may trust your first instinct. It really is trivial. Let us denote the box containing the two gold coins by B_{gg}, the box containing the two silver coins by B_{ss}, and the box containing one of each by B_{gs}. Since the boxes are identical, they have equal probabilities of being chosen. And since there is only one box in which the coins are different, we find the probability of having chosen B_{gs} to be $\frac{1}{3}$. I should mention that this question appeared on page 2 of the book and was intended merely to illustrate the idea of enumerating a set of equally likely possibilities.

Bertrand, however, did not leave things here. He went on to wonder how one ought to react to the following argument: Let us suppose we choose one box at random and remove one of the coins without looking at it. Regardless of the coin we choose, there are only two possibilities: the remaining coin in the box is either gold or it is silver. It is, therefore, either like or unlike the unexamined coin we have just removed. That makes two possibilities, each equally likely, and in only one of them are the coins different. It would seem that the removal of the coin caused our probability to jump from $\frac{1}{3}$ to $\frac{1}{2}$.

This argument is plainly fallacious, since the mere removal of one unidentified coin in no way increases our knowledge of the other coin in the box. Bertrand reasoned that the fallacy lay in assuming that the two possibilities (the coin being either like or unlike the coin we removed) were equally likely. In fact, since there are two boxes in which the coins are the same and only one in which they are different, it is self-evident that like coins are more probable than unlike coins. If we find, for example, that the coin we removed was gold, then the other coin is more likely to be gold than silver.

To see this, note that since the chosen coin is gold, we are removing from consideration the possibility that we chose B_{ss}. If we reached into B_{gg}, then the probability of removing a gold coin is equal to 1. But if we reached into B_{gs}, there is a probability of just $\frac{1}{2}$ of removing the gold coin. It follows that we are twice as likely to remove a gold coin having chosen B_{gg} than we would have been had we chosen B_{gs}. And since these probabilities must sum to 1, we find that the other coin will be silver with probability $\frac{1}{3}$ (and will be gold with probability $\frac{2}{3}$, just as we found previously).

Bertrand intended this as a cautionary tale of what happens when you are too cavalier in assigning equal probabilities to events. Lest you find this point too trivial to bother with, I assure you that some very competent mathematicians throughout history have managed to bungle it. In a famous example, the French mathematician Jean le Rond d'Alembert once argued that the probability of tossing at least one heads in two tosses of a coin is $\frac{2}{3}$. He argued there were only three possibilities: we could get a heads on the first toss or, barring that, get a tails on the first toss and then a heads on the second toss. We can represent these as H, TH, TT. He then treated these possibilities as equiprobable, from which his answer follows. Of course, a proper analysis would note that the coin comes up heads on the first toss half the time, while the scenario TH happens one-fourth of the time. Summing these possibilities leads to the correct probability of $\frac{3}{4}$ for getting heads at least once in two tosses of a coin.

A more modern form of Bertrand's problem begins with the same setup and asks for the probability that the second coin in the box is gold given that one coin was removed at random and seen to be gold. This gives us a standard problem in inverse probability. We initially assign an equal probability to each box. The new information is that our chosen box contains a gold coin. How ought we to revise our probability assessments? Bertrand's argument shows that updating the probability of an event given new information requires considering the probability of obtaining the information given the event. For example, the probability of having chosen B_{gs} given that we removed a gold coin depends in part on the probability of removing a gold coin having chosen B_{gs}. As we have mentioned, this insight lies at the heart of Bayes' theorem.

The similarity between this scenario and the Monty Hall problem is clear. In both scenarios an initial selection of three equiprobable options is narrowed to two in the light of new information. There is a tendency, in both cases, to assign an equal probability to the two remaining scenarios, but this must be resisted. Understanding Bertrand's problem is a useful first step toward resolving the Monty Hall scenario.

1.7. The Three Prisoners

In a 1959 column for *Scientific American*, Martin Gardner wrote:

> Charles Sanders Peirce once observed that in no other branch of mathematics is it so easy for experts to blunder as in probability theory. History bears this out. Leibniz thought it just as easy to throw 12 with a pair of dice as to throw 11. Jean le Rond d'Alembert, the great 18th century French mathematician, could not see that the results of tossing a coin three times are the same as tossing three coins at once, and he believed (as many amateur gamblers persist in believing) that after a long run of heads, a tail is more likely.

In light of the explosion over the Monty Hall problem that would occur just over three decades later, these words seem downright prophetic. That notwithstanding, our interest in this section resides in a particular brainteaser, presented by Gardner as follows:

A wonderfully confusing little problem involving three prisoners and a warden, even more difficult to state unambiguously, is now making the rounds. Three men—A, B and C—were in separate cells under sentence of death when the governor decided to pardon one of them. He wrote their names on three slips of paper, shook the slips in a hat, drew out one of them and telephoned the warden, requesting that the name of the lucky man be kept secret for several days. Rumor of this reached prisoner A. When the warden made his morning rounds, A tried to persuade the warden to tell him who had been pardoned. The warden refused.

"Then tell me," said A, "the name of one of the others who will be executed. If B is to be pardoned, give me C's name. If C is to be pardoned, give me B's name. And if I'm to be pardoned, flip a coin to decide whether to name B or C."

"But if you see me flip the coin," replied the wary warden, "you'll know that you're the one pardoned. And if you see that I don't flip a coin, you'll know it's either you or the person I don't name."

"Then don't tell me now," said A. "Tell me tomorrow morning."

The warden, who knew nothing about probability theory, thought it over that night and decided that if he followed the procedure suggested by A, it would give A no help whatever in estimating his survival chances. So next morning he told A that B was going to be executed.

After the warden left, A smiled to himself at the warden's stupidity. There were now only two equally probable elements in what mathematicians like to call the "sample space" of the problem. Either C would be pardoned or himself, so by all the laws of conditional probability, his chances of survival had gone up from $\frac{1}{3}$ to $\frac{1}{2}$.

The warden did not know that A could communicate with C, in an adjacent cell, by tapping in code on a water pipe. This A proceeded to do, explaining to C exactly what he had said to the warden and what the warden had said to him. C was equally overjoyed with the news because he figured, by the same reasoning used by A, that his own survival chances had also risen to $\frac{1}{2}$.

Did the two men reason correctly? If not, how should each calculate his chances of being pardoned?

This, surely, is the Monty Hall problem in all but name. Simply replace the three prisoners with three doors, the pardon with the car, the prisoners to be executed with the doors concealing the goats, and the warden with Monty Hall.

Now, I realize that it is a perilous thing for a historical researcher to declare that X is the first instance of Y. If it subsequently turns out that Y was lurking in some obscure corner of the academic literature, you can be certain that some overeducated braggart will delight in pointing out the fact. That risk notwithstanding, I would mention that I have dozens of professional references discussing this problem, and not one of them cites anything earlier than Gardner's column as its source. In personal correspondence Gardner was gracious enough to tell me he did not find the problem in any older published source, but rather heard the problem from various acquaintances. My own considerable researches have likewise failed to turn up any older reference. So I am calling it: Gardner's 1959 column is the first published instance of the Monty Hall problem, or at least of something formally equivalent to it.

Gardner presented the correct solution, that A will be pardoned with probability $\frac{1}{3}$ while C's chances have improved to $\frac{2}{3}$, in the following issue of the magazine [30]. He offered two arguments: first by enumerating the sample space, and alternatively by making an analogy to a situation with vastly more prisoners. Since the next chapter is devoted entirely to a consideration of such arguments, we will not discuss them here. We should also note the care with which Gardner stated the problem. In particular, he was explicit that the warden chooses randomly when given a choice of prisoners to name. This detail is essential to a proper solution of the problem, but it is often omitted in casual statements of it.

That said, in the spirit of showing just how difficult it can be to provide a truly pristine analysis of the problem, we can point to one unfortunate bit of phrasing in Gardner's presentation of the problem's solution. He writes, "Regardless of who is pardoned, the warden can give A the name of a man, other than A, who will die. The warden's statement therefore had no influence on A's survival chances; they continue to be $\frac{1}{3}$." Writing in [22], psychologist Ruma Falk points out the difficulty with this sentence: "Both parts of that sentence are correct, just the adverb 'therefore,' used here with conjunctive force, is inapt." The conclusion that A's probability does not change does not follow merely from the fact that the warden can always reveal the name of one of A's fellow prisoners. Rather, it is a consequence of the precise method used by the warden in deciding which name to reveal. We will revisit this point in the next chapter.

After Gardner's column, the three prisoners problem accumulated quite a literature, much of it providing an eerie parallel to the Monty Hall fracas that would erupt in the early nineties. Statistician Fred Mosteller included it as problem 13 in his *Fifty Challenging Problems in Probability with Solutions* [64] in 1965. In presenting the solution, he remarked that this problem attracts far more mail from readers than any other. Biologist John Maynard Smith, after presenting the problem in his 1968 book *Mathematical Ideas in Biology* [85], remarked, "This should be called the Serbelloni problem since it nearly wrecked a conference on theoretical biology in the summer of 1966; it yields at once to common sense or to Bayes' theorem." I certainly accept the latter part of that disjunction, but accumulated painful experience has left me dubious regarding the former.

As a case in point I would mention a statement made by Nicholas Falletta in his otherwise excellent 1983 book *The Paradoxicon* [23]. After presenting the three-prisoners problem, Falletta writes, "Prisoner *A* reasoned that since he was now certain that *B* would die then his chances for survival had improved from $\frac{1}{3}$ to $\frac{1}{2}$ and, indeed, they had!" Alas, Falletta did not explain how he came to this conclusion. He also neglected, in his statement of the problem, to tell his readers how the warden went about choosing which name to reveal. If he was envisioning the usual assumption, that the warden chooses randomly when given a choice, then we must regard Falletta's statement as simply incorrect. It is possible to imagine procedures the warden could follow that would justify Falletta's statement, but then these details needed to be spelled out. We will have more to say about this in the chapters ahead.

For now, let us note simply that not many math problems have been immortalized in verse. The prisoner's problem can claim that distinction, courtesy of mathematician Richard Bedient in [7].

The Prisoner's Paradox Revisited

Awaiting the dawn sat three prisoners wary
A trio of brigands named Tom, Dick and Mary
Sunrise would signal the death knoll of two
Just one would survive, the question was who.

Young Mary sat thinking and finally spoke
To the jailer she said, "You may think this a joke.
But it seems that my odds of surviving 'til tea,
Are clearly enough just one out of three.

But one of my cohorts must certainly go,
Without question, that's something I already know.
Telling the name of one who is lost,
Can't possibly help me. What could it cost?"

That shriveled old jailer himself was no dummy,
He thought, "But why not?" and pointed to Tommy.
"Now it's just Dick and I," Mary chortled with glee.
"One in two are my chances, and not one in three!"

Imagine the jailer's chagrin, that old elf.
She'd tricked him, or had she? Decide for yourself.

1.8. Let's Make a Deal

In 1963 the television game show *Let's Make a Deal* premiered on American television. In its initial run it lasted until 1977. In each episode the host, Monty

Hall, engaged in various games with members of his audience. These games had a number of formats, but the general principle was typically the same. Players had to decide between definitely winning a small prize and gambling on some probability of winning a greater prize.

In one game, which often served as the show's climax, contestants were shown three identical doors and were told that behind one of them was a car, while the other two concealed goats. Recall that in the abstract version of the problem considered here and in the next chapter, the game unfolds as follows: The contestant chooses but does not open a door. Monty now opens a door he knows to conceal a goat, choosing randomly when he has a choice. He then gives the player the options of sticking with his original choice or switching to the other unopened door. The contestant makes his choice and wins whatever is behind his door.

This is not, however, how things unfolded on the show. Typically, if the contestant initially chose a goat, the door was opened immediately and the game ended on the spot. But if the contestant chose the car, Monty opened one of the remaining doors and gave the contestant the option of switching. This option was sometimes accompanied by an offer of cash from Monty not to make the switch (if this offer was accepted, the player took the cash and went home, not opening any of the doors). If the player insisted on switching nonetheless, Monty would sometimes offer still more money, at times reaching as high as a few thousand dollars. This was a highly effective psychological ploy to make it seem that the car was behind the other remaining door. Of course, a devoted watcher of the show might have picked up on Monty's skullduggery, but that does not seem to have happened too often in practice.

Things do not get mathematically interesting until we stipulate that Monty always opens a goat-concealing door and always gives the option of switching. This might explain why it would be another dozen years until the term "Monty Hall problem" entered the mathematical literature.

1.9. The Birth of the Monty Hall Problem

In February 1975, the academic journal the *American Statistician* published a letter to the editor from Steve Selvin, then a mathematician at the University of California at Berkeley, proposing the following exercise in probability [80]. Given its considerable historical significance, we reproduce it in full:

> It is "Let's Make a Deal"—a famous TV show starring Monty Hall.
>
> MONTY HALL: One of the three boxes labeled *A*, *B* and *C* contains the keys to that new 1975 Lincoln Continental. The other two are empty. If you choose the box containing the keys, you win the car.
>
> CONTESTANT: Gasp!
>
> MONTY HALL: Select one of these boxes.
>
> CONTESTANT: I'll take box *B*.

MONTY HALL: Now box A and box C are on the table and here is box B (contestant grips box B tightly). It is possible the car keys are in that box! I'll give you $100 for the box.

CONTESTANT: No, thank you.

MONTY HALL: How about $200?

CONTESTANT: No!

AUDIENCE: No!!

MONTY HALL: Remember that the probability of your box containing the keys to the car is $\frac{1}{3}$ and the probability of your box being empty is $\frac{2}{3}$. I'll give you $500.

AUDIENCE: No!!

CONTESTANT: No, I think I'll keep this box.

MONTY HALL: I'll do you a favor and open one of the remaining boxes on the table (he opens box A). It's empty! (Audience: applause). Now either box C or your box B contains the car keys. Since there are two boxes left, the probability of your box containing the keys is now $\frac{1}{2}$. I'll give you $1000 cash for your box.

WAIT!!!!

Is Monty right? The contestant knows that at least one of the boxes on the table is empty. He now knows that it was box A. Does this knowledge change his probability of having the box containing the keys from $\frac{1}{3}$ to $\frac{1}{2}$? One of the boxes on the table has to be empty. Has Monty done the contestant a favor by showing him which of the two boxes was empty? Is the probability of winning the car $\frac{1}{2}$ or $\frac{1}{3}$?

CONTESTANT: I'll trade you my box B for the box C on the table.

MONTY HALL: That's weird!!

HINT: The contestant knows what he is doing.

The logic verifying the correctness of the contestant's strategy is then presented in the form of a table enumerating all the possibilities. It seemed straightforward enough, since a quick inspection revealed that in six out of nine possible scenarios, the contestant would win the car by switching. Considering the venue, a high-level journal read primarily by professional statisticians, you would have expected a raised eyebrow or two and little more. But this is the Monty Hall problem we are discussing, and it has the power to make otherwise intelligent people take leave of their senses.

Selvin's letter was published in February. By August, Professor Selvin was back in the letters page with a follow-up [81]. His letter was entitled "On the Monty Hall Problem," and we note with great fanfare this earliest known occurrence of that phrase in print. Selvin noted that he received a number of letters in response to his earlier essay taking issue with his proposed solution. He went on to present a second argument in defense of his conclusion, this time a more technical one involving certain formulas from conditional probability.

Especially noteworthy is the following statement from Selvin's follow-up: "The basis to my solution is that Monty Hall knows which box contains the keys and when he can open either of two boxes without exposing the keys, he chooses between them at random." In writing this he had successfully placed his finger on the two central points of the problem. Alter either of those assumptions and the analysis can become even more complex, as we shall see in the chapters to come.

In an amusing coda to this story, Selvin notes in his follow-up that he had received a letter from Monty Hall himself:

> Monty Hall wrote and expressed that he was not a "student of statistics problems" but "the big hole in your argument is that once the first box is seen to be empty, the contestant cannot exchange his box." He continues to say, "Oh, and incidentally, after one [box] is seen to be empty, his chances are no longer 50/50 but remain what they were in the first place, one out of three. It just seems to the contestant that one box having been eliminated, he stands a better chance. Not so."

It would seem that Monty Hall was on top of the mathematical issues raised by his show. It is a pity that more mathematicians were not aware of Selvin's lucid analysis. They might thereby have spared themselves considerable public embarrassment.

Another decade and a half would go by before the Monty Hall problem really left its mark on the mathematical community. While the three-prisoners problem continued to feature prominently in professional articles from various disciplines, for example in a 1986 paper [20] by Persi Diaconis and Sandy Zabell discussing approaches to inverse probability different from Bayes' theorem, the Monty Hall problem was mostly dormant throughout the eighties. It popped up in 1987, when Barry Nalebuff [66] presented it in the inaugural problem section of the academic journal *Economic Perspectives*. For the most part, however, the problem was still flying decidedly below the radar during this time.

1.10. *L'Affaire* Parade

It is September 9, 1990. President Samuel Doe of the small African nation of Liberia is assassinated by rebel forces as part of one of the bloodiest Civil Wars that continent would ever see. United States president George H. W. Bush and Russian president Mikhail Gorbachev present a joint statement protesting the illegal occupation of Kuwait by Iraqi military forces. Tennis star Pete Sampras won the first of his record-setting fourteen Grand Slam tennis championships by defeating fellow American Andre Agassi in the finals of the U.S. Open. The uncut version of Stephen King's horror masterpiece *The Stand* rests at number five on the *New York Times* best-seller list.

And Marilyn vos Savant, a Q & A columnist for *Parade* magazine, responds to the following question from reader Craig Whitaker of Columbia, Maryland [95]:

Suppose you're on a game show, and you're given the choice of three doors. Behind one door is a car, behind the others, goats. You pick a door, say number 1, and the host, who knows what's behind the doors, opens another door, say number 3, which has a goat. He says to you, "Do you want to pick door number 2?" Is it to your advantage to switch your choice of doors?

Ms. vos Savant replied as follows:

Yes, you should switch. The first door has a $\frac{1}{3}$ chance of winning, but the second door has a $\frac{2}{3}$ chance. Here's a good way to visualize what happened. Suppose there are a *million* doors, and you pick door number 1. Then the host, who knows what's behind the doors and will always avoid the one with the prize, opens them all except door number 777,777. You'd switch to that door pretty fast, wouldn't you?

With this exchange we open one of the strangest chapters in the history of mathematics.

The Oxford University biologist Richard Dawkins once responded to an extremely hostile, and badly misinformed, review of one of his books [18] by writing, "Some colleagues have advised me that such transparent spite is best ignored, but others warn that the venomous tone of her article may conceal the errors in its content. Indeed, we are in danger of assuming that nobody would dare to be so rude without taking the elementary precaution of being right in what she said." He might as well have been discussing the response to vos Savant's proposed solution.

Mind you, I can understand why someone, even a professional mathematician, would be caught out by the Monty Hall problem. It is genuinely counterintuitive, even for people with serious training in probability and statistics. As we shall see in the next chapter, no less a personage than Paul Erdös, one of the most famous mathematicians of the twentieth century, not only got the problem wrong but stubbornly refused to accept the correct answer for quite some time. The prominent Stanford University mathematician Persi Diaconis once said of the Monty Hall problem [89], "I can't remember what my first reaction to it was because I've known about it for so many years. I'm one of many people who have written papers about it. But I do know that my first reaction has been wrong time after time on similar problems. Our brains are just not wired to do probability problems very well, so I'm not surprised there were mistakes."

But if getting it wrong is understandable, being snotty and condescending about it is not. In a follow-up column on December 2, vos Savant shared some of the choicer items from her mailbox. I have not troubled here to reproduce the names of the correspondents, since it is not my intention to embarrass anyone. The morbidly curious can check out vos Savant's book [95]. Suffice it to say that the correspondents below were all mathematicians:

> Since you seem to enjoy coming straight to the point, I'll do the same. In the following question and answer, you blew it! Let me explain. If one door is shown to be a loser, that information changes the probability of either remaining choice, *neither of which has any reason to be more likely*, to $\frac{1}{2}$. As a professional mathematician, I'm very concerned with the general public's lack of mathematical skills. Please help by confessing your error and in the future being more careful.

And:

> You blew it, and you blew it big! Since you seem to have difficulty grasping the basic principle at work here, I'll explain. After the host reveals a goat, you now have a one-in-two chance of being correct. Whether you change your selection or not, the chances are the same. There is enough mathematical illiteracy in this country, and we don't need the world's highest I.Q. propagating more. Shame!

And:

> Your answer to the question is in error. But if it is any consolation, many of my academic colleagues have also been stumped by this problem.

In replying, vos Savant rightly stuck to her original answer. This time she opted for the approach of enumerating the sample space, which really ought to have ended the discussion. It did not.

On February 17, 1991, vos Savant revisited the problem yet again. Once more she shared the musings of some of her more obnoxious correspondents:

> May I suggest that you obtain and refer to a standard textbook on probability before you try to answer a question of this type again?

And:

> I have been a faithful reader of your column, and I have not, until now, had any reason to doubt you. However, in this matter (for which I do have expertise), your answer is clearly at odds with the truth.

And:

> You are utterly incorrect about the game-show question, and I hope this controversy will call some public national attention to the serious national crisis in mathematical education. If you can admit your

error, you will have contributed constructively towards the solution of a deplorable situation. How many irate mathematicians are needed to get you to change your mind?

And:

You made a mistake, but look at the positive side. If all those PhD's were wrong, the country would be in some very serious trouble.

Looks like the country is in serious trouble.

Marilyn vos Savant's answer this time introduced two important nuances into the discussion. First, she stated more explicitly than previously that it is crucial to assume that Monty always opens a losing door. Relaxing that assumption in any way changes the problem completely.

Second, she proposed that mathematics classes across the country put her proposed solution to the test. The Monty Hall scenario is easily simulated, making it possible to run through a large number of trials in a relatively short period of time. By having one segment of the class follow an "always switch" strategy while having the other electing to "always stick," it becomes a straightforward matter to see who wins more frequently. In mathspeak we would say that vos Savant was proposing the question be resolved via a Monte Carlo simulation.

Many folks took her up on this suggestion, leading to a fourth column on the subject. The letters inspiring this one were of a considerably different tone:

In a recent column, you called on math classes around the country to perform an experiment that would confirm your response to a game show problem. My eighth-grade classes tried it, and I don't really understand how to set up an equation for your theory, but it definitely does work. You'll have to help rewrite the chapters on probability.

And:

Our class, with unbridled enthusiasm, is proud to announce that our data support your position. Thank you so much for your faith in America's educators to solve this.

And:

I must admit I doubted you until my fifth-grade math class proved you right. All I can say is wow!

And:

After considerable discussion and vacillation here at the Los Alamos National Laboratory, two of my colleagues independently programmed the problem, and in 1 million trials, switching paid off 66.7 percent of the time. The total running time on the computer was less than one second.

Most amusing of all, one letter writer had the audacity to write: "Now 'fess up. Did you really figure all this out, or did you get help from a mathematician?"

So vos Savant was vindicated. Nothing succeeds like success, and to anyone actually playing the game multiple times it quickly becomes clear that switching is the way to go.

This story has a curious footnote. Shortly after vos Savant wrote her last column on the Monty Hall problem, she received the following letter:

> A shopkeeper says she has two new baby beagles to show you, but she doesn't know whether they're male, female, or a pair. You tell her that you want only a male, and she telephones the fellow who's giving them a bath. "Is at least one a male?" she asks him. "Yes!" she informs you with a smile. What is the probability that the *other* one is a male?

She replied with the correct answer that the probability is one out of three. There are three ways to have a pair of puppies in which one is male, you see, and these scenarios are equally likely. Listing the puppies in the order of their birth, they could be male/female, female/male or male/male. Since it is only the last of these scenarios in which "the other puppy" is male, we arrive at our answer. This problem, in various forms, is itself a classic problem in probability, and we shall have more to say about it in Chapter 6.

By now you should not be surprised to learn that a storm of angry correspondence ensued, most of them lecturing vos Savant about how the sex of each puppy is entirely independent of the sex of the other one. A family could have five sons in a row, but the chances are still fifty-fifty that the next child will be a daughter.

This is not in dispute. It also is not what was asked. Had the problem said, "The older puppy is a male. What is the probability that the younger puppy is a male also?" then the answer would surely be $\frac{1}{2}$. As it is, however, there is no first puppy or second puppy specified in the problem.

1.11. The *American Statistician* Exchange

The humiliation dealt to the mathematical community in the wake of *l'affaire Parade* could not be ignored. Countless simulations made clear the fact that Marilyn vos Savant had answered the question correctly, suggesting that her rather intemperate correspondents had a lot of crow to eat. That notwithstanding, it was possible that some portion of the blame could still be laid at her feet. While her answer was surely correct, perhaps her reasoning left something to be desired. This tactic took its most pointed form in an exchange of letters in the academic journal the *American Statistician* [60], [61].

I have already reproduced vos Savant's first solution to the problem. Let us now consider her subsequent attempts. Her second gambit was the following:

My original answer is correct. But first, let me explain why your answer is wrong. The winning chances of $\frac{1}{3}$ on the first choice can't go up to $\frac{1}{2}$ just because the host opens a losing door. To illustrate this, let's say we play a shell game. You look away, and I put a pea under one of three shells. Then I ask you to put your finger on a shell. The chances that your choice contains a pea are $\frac{1}{3}$, agreed? Then I simply lift up an empty shell from the remaining two. As I can (and will) do this regardless of what you've chosen, we've learned nothing to allow us to revise the chances on the shell under your finger.

The benefits of switching are readily proven by playing through the six games that exhaust all the possibilities. For the first three games, you chose number 1 and "switch" each time, for the second three games, you choose number 1 and "stay" each time, and the host always opens a loser. Here are the results:

She now produced a table listing the various scenarios. In the interests of conserving space, I will note that she listed three scenarios which can be described as *AGG*, *GAG*, and *GGA*, depending on the location of the car. These scenarios are equally likely. If we now assume that you always choose door one and that Monty only opens goat-concealing doors, we see than in two of the three scenarios you win by switching. She then concluded:

When you switch, you win $\frac{2}{3}$ of the time and lose $\frac{1}{3}$, but when you don't switch, you only win $\frac{1}{3}$ of the time and lose $\frac{2}{3}$. You can try it yourself and see.

Alternatively, you can actually play the game with another person acting as the host with three playing cards—two jokers for the goats and an ace for the prize. However, doing this a few hundred times to get statistically valid results can get a little tedious, so perhaps you can assign it as extra credit—or for punishment! (*That'll* get their goats!)

Vos Savant gave a still more detailed treatment in her third column. I will beg your indulgence as I present a lengthy excerpt. It is necessary for fully understanding what happened next.

So let's look at it again, remembering that the original answer defines certain conditions, the most significant of which is that *the host always opens a losing door on purpose*. (There's no way he can always open a losing door by chance!) Anything else is a different question.

The original answer is still correct, and the key to it lies in the question *"Should you switch?"* Suppose we pause at this point, and a UFO settles down onto the stage. A little green woman emerges, and the host asks her to point to one of the two unopened doors. The chances that *she'll* randomly choose the one with the prize are $\frac{1}{2}$, all right. But that's because she lacks an advantage the *original* contestant had—the help of the host. (Try to forget any particular television show.)

When you first choose door number 1 from three, there's a $\frac{1}{3}$ chance that the prize is behind that one and a $\frac{2}{3}$ chance that it's behind one of the others. *But then the host steps in and gives you a clue.* If the prize is behind number 2, the host shows you number 3, and if the prize is behind number 3, the host shows you number 2. So when you switch, you win if the prize is behind number 2 *or* number 3. *You win either way!* But if you *don't* switch, you win only if the prize is behind door number 1.

And as this problem is of such intense interest, I'll put my thinking to the test with a nationwide experiment. This is a call to math classes all across the country. Set up a probability trial exactly as outlined below and send me a chart of all the games along with a cover letter repeating just how you did it, so we can make sure the methods are consistent.

One student plays the contestant, and another, the host. Label three paper cups number 1, number 2, and number 3. While the contestant looks away, the host randomly hides a penny under a cup by throwing a die until a one, two, or three comes up. Next, the contestant randomly points to a cup by throwing a die the same way. Then the host purposely lifts up a losing cup from the two unchosen. Lastly, the contestant "stays" and lifts up his original cup to see if it covers the penny. Play "not switching" two hundred times and keep track of how often the contestant wins.

Then test the other strategy. Play the game the same way until the last instruction, at which point the contestant instead "switches" and lifts up the cup *not* chosen by anyone to see if it covers the penny. Play "switching" two hundred times, also.

Certainly vos Savant's arguments are not mathematically rigorous, and we can surely point to places where her phrasing might have been somewhat more precise. Her initial argument based on the million-door case is pedagogically effective but mathematically incomplete (as we shall see). And there was a subtle shift from the correspondent's initial question, in which the host always opens door three, to the listing of the scenarios given by vos Savant, in which it was assumed only that the host always opens a goat-concealing door.

But for all of that, it seems clear that vos Savant successfully apprehended all of the major points of the problem and explained them rather well considering the forum in which she was writing. Her intent was not to provide an argument of the sort a mathematician would regard as definitive, but rather to illuminate the main points at issue with arguments that would be persuasive and comprehensible. In this she was successful.

Four people who were less impressed were mathematicians J. P. Morgan, N. R. Chaganty, R. C. Dahiya, and M. J. Doviak (collectively MCDD). Writing in the *American Statistician* [61], they presumed to lay down the law regarding vos Savant's treatment of the problem. After quoting the original question as posed by vos Savant's correspondent, they write:

> Marilyn vos Savant, the column author and reportedly holder of the world's highest I.Q., replied in the September article, "Yes, you should switch. The first door has a $\frac{1}{3}$ chance of winning, but the second door has a $\frac{2}{3}$ chance." She then went on to give a dubious analogy to explain the choice. In the December article letters from three PhD's appeared saying that vos Savant's answer was wrong, two of the letters claiming that the correct probability of winning with either remaining door is $\frac{1}{2}$. Ms. vos Savant went on to defend her original claim with a false proof and also suggested a false simulation as a method of empirical verification. By the February article a full scale furor had erupted; vos Savant reported. "I'm receiving thousands of letters nearly all insisting I'm wrong. ... Of the letters from the general public, 92% are against my answer; and of letters from universities, 65% are against my answer." Nevertheless, vos Savant does not back down, and for good reason, as, given a certain assumption, her answer is correct. Her methods of proof, however, are not.

Rather strongly worded, wouldn't you say? And largely unfair, for reasons I have already discussed. Indeed, continuing with their lengthy essay makes clear that their primary issue with vos Savant is her shift from what they call the "conditional problem," as posed by her correspondent (in which it is stipulated that the contestant always chooses door one and the host always opens door three), to the "unconditional problem," in which we stipulate only that after the contestant chooses a door, the host opens one of the goat-concealing doors. She did, indeed, make this shift, but this was hardly the point at issue between vos Savant and her angry letter writers. (We will revisit the distinction between the conditional and the unconditional problem in Chapter 7.)

The authors go on to provide an interesting and useful probabilistic analysis of the problem, but that is not our interest here. At present we are interested in the human drama that surrounded the appearance of the problem in vos Savant's column. With that in mind, let us ponder vos Savant's response to MCDD. After quoting the original question and her original answer, vos Savant writes,

> It should be understood by an academic audience that this problem, written for a popular audience, was not intended to be subject to strong

attempts at misinterpretation. If it had been, it would have been a page long. While it may be instructionally constructive to purposely focus on semantic issues here, it is surely intellectually destructive to imply that it reflects negatively on the perspicacity of the writer involved.

And later,

Nearly all of my critics understood the intended scenario, and few raised questions of ambiguity. I personally read nearly three thousand letters (out of many more thousands that ultimately arrived) and found virtually every reader, from university lectern to kitchen table, insisting simply that because two options remained, the chances were even.

I would continue, but I find my initial annoyance flagging and will instead devote my energy to confounding the editorial staff at *Parade* once again, especially as it has occurred to me that the authors have clearly at least found a way to provoke me to sit down and write a response when other readers have failed to do so. Frankly, after seeing this problem analyzed on the front page of the *New York Times* and now creating a similar stir in England, I have given up on getting the facts across properly and have decided simply to sit back and amuse myself with the reading of it all.

Zing!

MCDD responded to this. They had the audacity to begin with, "We are surprised at the tone of vos Savant's reply." It is unclear what tone they were expecting in response to their bellicose and condescending essay. They then repeated the main points from their earlier essay, emphasizing that the problem vos Savant discussed was not precisely the problem laid out in the initial question. This point is not at issue, but what vos Savant discussed was surely what was intended. Even if it was not, vos Savant made it quite clear what problem she *was* discussing. Seen in that light. MCDD ought not to have said that her arguments were wrong and contained technical errors when they meant simply that she had altered the problem slightly from what was originally stated. In fairness, MCDD do moderate their tone later on, writing. "None of this diminishes the fact that vos Savant has shown excellent probabilistic judgment in arriving at the answer $\frac{2}{3}$, where, to judge from the letters in her column, even member of our own profession failed."

I have belabored this incident for two reasons. The first is to capture for you the heat and emotion that has characterized so many discussions of this issue, even in otherwise staid, professional outlets. I would hardly be doing my job as a chronicler of all things related to the Monty Hall problem if I did otherwise.

The other is to illustrate what I perceive as an occupational hazard among mathematicians—specifically, the desire always to be the smartest person in the room. In my experience, this sad tendency is especially prevalent when interacting with non-mathematicians. The relish with which MCDD declare vos Savant's arguments to be wrong is both palpable and completely uncalled-for.

They, at least, were mathematically correct in their substantive points. The motives of the letter writers whose hectoring and arrogant missives have deservedly earned them a place in the mathematical hall of shame are even more incomprehensible. What could possibly make people think it is acceptable to write with such a tone over a mere exercise in probability theory?

1.12. The Aftermath

Writing in the magazine *Bostonia* [73], cognitive scientist Massimo Piatelli-Palmarini aptly described the Monty Hall problem by writing, "No other statistical puzzle comes so close to fooling all the people all the time.... The phenomenon is particularly interesting precisely because of its specificity, its reproducibility, and its immunity to higher education."

In a front-page article for the Sunday *New York Times* on July 21, 1991, John Tierney summed things up as follows:

> Since she gave her answer, Ms. vos Savant estimates she has received 10,000 letters, the great majority disagreeing with her. The most vehement criticism has come from mathematicians and scientists, who have alternated between gloating at her ("You are the goat!") and lamenting the nation's innumeracy.
>
> Her answer—that the contestant should switch doors—has been debated in the halls of the Central Intelligence Agency and the barracks of fighter pilots in the Persian Gulf. It has been analyzed by mathematicians at the Massachusetts Institute of Technology and computer programmers at Los Alamos National Laboratory in New Mexico. It has been tested in classes from second grade to graduate level at more than 1,000 schools across the country.

Since the initial fracas erupted, the Monty Hall problem has accumulated a formidable technical literature. It would seem that researchers from a wide variety of disciplines found something of interest within its simple scenario. Mathematicians and statisticians hashed out the probabilistic issues raised by the problem and its variants [3; 10; 14; 33; 54; 77; 79; 82]. Philosophers found connections between the Monty Hall problem and various long-standing problems in their own discipline [5; 6; 11; 15; 44; 53; 62]. Physicists devised quantum mechanical versions of the problem [16; 101]. Cognitive scientists and psychologists tried to determine why, exactly, people have so much trouble with this problem [1; 34; 36; 37; 38; 39; 45; 46; 52]. Economists pondered the relevance of the Monty Hall problem to the problems of human decision making [48; 72; 74; 84; 92]. These are just a few representative citations. There are many others.

We will have occasion to look at much of this research in the pages ahead, but this introduction has gone on long enough. It is time to do some math!

1.13. Appendix: Dignity in Problem Statements

Steve Selvin gave the first-ever published presentation of the Monty Hall problem in the form of a play. In doing so he inaugurated a disturbing trend in published statements of the problem. Apparently believing the scenario is insufficiently confusing when presented flat-out as a teaser in probability, many authors feel the need to embed it within some larger bit of melodrama. Here is an example, taken from [58] (all emphases in original):

> ANNOUNCER: And now ... the game show that mathematicians argue about ... LET"S MAKE A DEAL. Here's your genial host, Monty Hall! [Applause]
>
> MONTY: Hello, good evening, and welcome! Now let's bring up our first contestant. It's ... YOU! Come right up here. Now, you know our rules. Here are three doors, numbered 1, 2, and 3. Behind one of these doors is a beautiful new PONTIAC GRAN HORMONISMO!
>
> AUDIENCE: Oooh! Aahh!
>
> MONTY: Behind the other two are WORTHLESS GOATS!
>
> AUDIENCE: [Laughter]
>
> GOATS: Baah!
>
> MONTY: Now, you're going to choose one of those doors. Then I'm going to open one of the other doors with a goat behind it, and show you the goat. Then I'll offer you this deal: if you stick with the door you've chosen, you can keep what's behind it, plus $100. If instead you chose the remaining unopened door, you can keep what's behind it. Now choose one door.
>
> AUDIENCE: Pick 3! No, 1! 2!
>
> YOU: Um, oh well, I guess I'll pick ... 3.
>
> MONTY: Okay. Now our beautiful host Charleen will open door number 2. Inside that door, as you can see, is a WORTHLESS GOAT. You can keep what's behind your door 3 plus $100, or you can make a deal and switch for whatever's behind door 1. While we take our commercial break, you should decide: do you wanna MAKE A DEAL??

It's the line where the goats say "Baah!" that really makes you feel like you're there.

In the course of describing the relevance of the Monty Hall problem to bridge players [57] (details in Chapter 5), Phil Martin serves up the following presentation:

> "Behind one of these three doors," shouts Monty Hall, "is the grand prize, worth one hundred thousand dollars. It's all yours—if you pick the right door."
>
> "I'll take door number one," you say.
>
> "Let's see what's behind door number—No! Wait a minute!" says Monty Hall. "Before we look, I'll offer you *twenty thousand dollars*, sight unseen, for whatever's behind door number one."
>
> "No! No!" shouts the audience.
>
> "Of course not," you say. "Even assuming the booby prizes are worth nothing, the expected value of my choice is thirty-three and a third thousand dollars. Why should I take twenty thousand?"
>
> "All right, says Monty Hall. "But before we see what you've won, let's take a look behind *door number two!*"
>
> Door number two opens to reveal one of the booby prizes: a date in the National Open Pairs with Phil Martin. You and the audience breathe a sigh of relief.
>
> "I'll give you one last chance," says Monty Hall. "You can have *forty* thousand dollars for what's behind door number one."
>
> "No, no!" shouts the audience.
>
> "Sure," you say.

I really must protest these cheap theatrics. The problem has all it can handle getting itself stated with sufficient clarity to be mathematically tractable. Embedding it in a skit only makes it harder to parse, and typically, as in the two examples above, leads to important assumptions not being spelled out. So knock it off!

2

Classical Monty

We begin with some terminological ground rules. The host of our hypothetical game show shall always be referred to as "Monty." We will use the second person in describing the game-show contestant. That is, we will phrase things as if you, the reader, are playing the game. A door will be said to be "correct" if it conceals the car and "incorrect" or "empty" if it conceals a goat. When we speak of being presented with identical doors, you should interpret that to mean that each of the doors has an equal probability of being correct. Your goal in every case is to maximize your chances of winning the car. If it is said that Monty makes a decision at random, that means he chooses from among his possibilities with equal probability.

The time has come to make precise the notions discussed in the previous chapter. Let us have a look.

2.1. The Canonical Version

Version One: You are shown three identical doors. Behind one of them is a car. The other two conceal goats. You are asked to choose, but not open, one of the doors. After doing so, Monty, who knows where the car is, opens one of the two remaining doors. He always opens a door he knows to be incorrect, and randomly chooses which door to open when he has more than one option (which happens on those occasions where your initial choice conceals the car). After opening an incorrect door, Monty gives you the option of either switching to the other unopened door or sticking with your original choice. You then receive whatever is behind the door you choose. What should you do?

When I was five years old my father asked me the following question: "If you are standing on the Earth and you want to see the Moon, what should you do?" "Look up!" I replied. "Very good," said my father. "Now, suppose you are standing on the Moon and want to see the Earth. What should you do?" A five-year-old's logic being what it is, I gave the obvious answer: "Look down!" My father smiled at me. "But if you look down," he said, "you will see your feet."

I have no recollection of these events, but my father assures me that my subsequent look of befuddlement was a sight to behold. Two entirely reasonable arguments that nonetheless pointed to different conclusions. On the one hand, if you look down you will see your feet. Hard to argue with that, even at age five. On the other hand, if from Point A you must look up to see Point B, then it stands to reason that from Point B you should look down to see Point A. Both are plausible arguments, but one of them simply must be wrong.

So it is with the Monty Hall problem. Upon hearing it for the first time, most people respond as follows: "After Monty opens a door there are only two options remaining. As far as I know, these options are equally likely. There is therefore a probability of $\frac{1}{2}$ that either particular door is correct. Given that, it makes no difference whether or not I switch doors."

This seems plausible until someone asks you what you would do if you started with one hundred identical doors. One conceals a valuable prize, while the other ninety-nine conceal nothing of value. You choose one at random, after which Monty randomly opens ninety-eight doors he knows to be empty. He then gives you the option of switching to the one remaining door. Do you really want to argue that the door you chose randomly from a hundred identical possibilities is as likely to be correct as that one other unopened door?

Surely not. That notwithstanding, the manifest absurdity of not switching in the hundred-door scenario does nothing to make our initial argument seem absurd in the three-door scenario. The question remains: what should we do to maximize our chances of winning? We have two plausible arguments that point to contradictory conclusions. One of them must be wrong, but which one?

Answering that question requires some discussion of probability.

We should, however, make one further point before getting on with that grim piece of business. I have followed tradition in presenting the Monty Hall problem as an exercise in decision theory. That is, we seek the most rational decision for the contestant to make given the available information. When we speak of solving the problem, however, we will have something slightly different in mind. We will regard the problem as solved when we can assign the correct probability to each door at every stage of the game. If we are successful, then we will surely be able to determine the most rational course of action. In principle, however, something less than this could be adequate for answering the narrow question of the most rational course of action. If you are inclined to argue, "Well, either the doors are equally likely or the unchosen door is more likely, so I might help myself and cannot hurt myself by switching, and that is what I should do," then I will give you a grade of incomplete.

2.2. Arguments

Let us consider what can be said in defense of our various options.

2.2.1. You Should Switch!

In my experience, the multiple-door argument for switching given in the previous section is convincing to almost everyone. That is, we consider the case of one hundred identical doors. Your initial choice is correct with probability $\frac{1}{100}$. That means that in 99 out of 100 cases, the prize is behind one of the other 99 doors. And since Monty only opens empty doors, we conclude that in nearly every case the prize will be behind the one other remaining door. Therefore, you should switch. We saw in the previous chapter that Marilyn vos Savant made this her lead argument in discussing the problem.

Students who are totally unpersuaded by elaborate probability calculations or arguments based on Bayes' theorem typically cry uncle at this point. But not always. I have had many people tell me, with perfect sincerity, that by opening ninety-eight empty doors Monty has caused the probability that your initial choice is correct to increase from $\frac{1}{100}$ to $\frac{1}{2}$. This argument is not as foolish as it looks. As we shall see in Chapter 3, in certain variants of the problem Monty's actions really do alter the probability of winning with your initial choice.

Another retort, suggested to me by an enthusiastic young statistics major, is that the two scenarios are not comparable. Sure, in the hundred-door case you should switch. But logic that applies to such a wealth of doors does not necessarily hold for a much smaller space. I regret to say I never did manage to talk her out of this view.

The gravest difficulty with this argument is simply that it is incomplete as it stands. The assumption that Monty only opens empty doors does not, by itself, justify the conclusion that the probability of your initial choice remains unchanged. As we shall see, the procedure used by Monty in choosing a door to open must also be taken into consideration.

Here is a closely related argument, taken from [58]. The problem had been stated under the assumption that you choose door three initially.

The probability that door 3 has the car is $\frac{1}{3}$; so the probability that it doesn't—that it's behind 1 or 2—is $\frac{2}{3}$. After Monty opens one of these, which he knows hides a goat, the probability of the other hiding a car is now $\frac{2}{3}$.

This argument is sadly ubiquitous in casual treatments of the problem. It is, like the multiple-door argument considered previously, incomplete as it stands. It is simply not true in general that if n equiprobable events have a collective probability of p and then information is received that decreases the probability of one of those events to 0, then the remaining events should still be viewed as equiprobable with a collective probability of p. In the present situation, we have that prior to Monty's opening a door, doors one and two are equiprobable and have probabilities summing to $\frac{2}{3}$. It does not follow from this that when Monty opens one of the doors, the entire $\frac{2}{3}$ probability then shifts to the one remaining door. More is needed to justify this conclusion.

Another casual, but incomplete, argument asserts that nothing Monty did in opening an empty door alters the fact that your initial choice is only correct one-third of the time. Since switching converts a win to a loss or a loss to a win, you will win two-thirds of the time by switching.

Sadly, this merely assumes what we are trying to prove. After all, the whole point at issue is whether Monty's actions give us cause to alter our assessments of the probabilities of winning with various doors. Our initial belief that our door is correct with probability $\frac{1}{3}$ was based on the premise of confronting three identical doors. After Monty opens an incorrect door, that situation no longer holds. We have received new information that may, or may not, give us reason to alter our assessments of the probabilities. So we cannot merely assert that Monty's actions fail to alter the probabilities. That must be proved.

Just so we are clear, the conclusion of these arguments is correct. You really do double your chances by switching. It is simply that we currently lack a rigorous argument for that finding.

2.2.2. It Doesn't Matter!

Upon hearing the problem the first time, most people reply as previously indicated. After Monty opens a door, there are two, equally likely, doors remaining. Thus, there is no advantage to be gained from switching. It is not so straightforward to provide an intuitive reason for discarding this argument.

The logic here essentially ignores the classical version of the problem and replaces it with something different. The suggestion is that after Monty eliminates a door, we should act as if we are confronted with a new problem in which only two identical doors are provided. In such a situation we plainly have no reason to prefer one over the other. Were someone to walk into the theater at that moment, oblivious to the prior history of the game, she would see two identical doors and have no basis for choosing between them. Indeed. But is your situation identical in all relevant ways to that of our newcomer?

Arguing in this way, you see, ignores the fact that Monty operates under certain restrictions in deciding which door to open. Specifically, he is not allowed to open the door you have chosen, and be is not allowed to open the door with the car. He is also required to choose his door randomly in those cases where he has more than one option (that is, in situations where your initial choice conceals the car). When you see Monty open, say, door two, the information you have received is not merely "There is a goat behind door two," but rather "Monty, who makes his decisions according to known rules, chose to reveal that there is a goat behind door two." If your initial choice was door one, then the event "Monty opens door two" is more likely to happen if the car is behind door three (since if you chose door one and the car is behind door three, Monty is *forced* to open door two) than if it is behind door one (since now Monty must open either door two or door three, and he will choose randomly between them). Thus, if we now see Monty open door two, we should conclude that it is more likely that the car is behind door three than behind door one.

If you accept this logic, you might be wondering what happens if Monty is not guaranteed to open an empty door. For example, what if he simply chooses a door at random and lucks into choosing an empty one? This possibility is the subject of Chapter 3. As we shall see, in this case it does not matter which door you choose, since now each remaining door really does have a probability of $\frac{1}{2}$.

A more fanciful defense of this position was provided by Raymond Smullyan in his collection of logic puzzles *The Riddle of Scheherazade and Other Amazing Puzzles* [86]. Allow me to turn the floor over to him:

"And now," said Scheherazade, "I have a paradox for you. There are three boxes labeled A, B and C. One and only one of the three boxes contains a gold coin; the other two are empty. I will prove to you that regardless of which of the three boxes you pick, the probability that it contains the gold coin is one in two."

"That's ridiculous!" said the king. "Since there are three boxes the probability is clearly one in three."

"Of course it's ridiculous," said Scheherazade, "and that's what makes it a paradox. I will give you proof that the probability is one in two, and your problem is to find the error in the proof, since the proof must obviously contain an error."

"All right," said the king.

"Let's suppose you pick Box A. Now, the coin is with equal probability in any of the three boxes, so if Box B should be empty, then the chances are fifty-fifty that the coin is in Box A."

"Right," said the king.

"Also, if Box C is empty, then again the chances are fifty-fifty that the coin is in Box A."

"That's right," said the king.

"But at least one of the boxes, B or C, must be empty, and whichever one is empty, the chances are fifty-fifty that the coin is in Box A. Therefore the chances are fifty-fifty, period!"

"Oh, my!" said the king.

As is clear from the excerpt above, Smullyan intended this simply as an exercise, not as a serious argument. Still, it is instructive to pinpoint precisely where things go wrong. The argument bears a striking resemblance to Bertrand's argument, given in section 1.6.

The problem is simply that the conclusion does not follow from the premises. It is certainly correct to say that if we consider only those cases where the coin is not in box B, then the coin will be in box A half the time. It is also correct to say that if we consider only those cases where the coin is not in box C, then the coin will be in box A half the time. The fact remains, however, that we are interested only in the fraction of cases in which the coin is neither in box B nor in box C. And *that* happens two-thirds of the time, our previous statements notwithstanding.

There is a parallel here with the two-ace problem considered in the previous chapter. There we considered two standard, well-shuffled decks of fifty-two cards. We sought the probability of having the ace of spades appear as the top card on at least one of the decks. Our first argument involved treating the decks as two separate entities. Since there is a probability of $\frac{1}{52}$ of obtaining the ace on either deck by itself, we erroneously claimed there is a probability of $\frac{1}{26}$ of getting the ace on at least one of the two decks.

The error came in failing to consider the possibility of obtaining the ace on both decks simultaneously. It is not correct to treat the decks as entirely separate entities, because there is some overlap between the instances in which the first deck provides the ace with the instances in which the second deck does so. So it is with Smullyan's argument about the boxes. In listing the possible scenarios, there is considerable overlap between the cases where box B is empty with the cases where box C is empty. This overlap makes it illegitimate to treat these scenarios as seaprate affairs.

2.2.3. Do Not Switch!

While the Supporters of Switching are slugging it out with the Armies of Indifference, we should be careful not to ignore the arguments to be made for sticking with your original door. Numerous experiments have shown that sticking is, by a considerable margin, the most popular choice. This presents a bit of a puzzle. I am not aware of any attempted mathematical argument that shows an actual advantage to sticking; the usual conclusion is simply that the two remaining doors are equally likely to be correct. In that case, you might expect at least some people to switch just for the sheer capriciousness of it all. That, however, does not seem to be the case.

My ever-resourceful students have provided me with two arguments in defense of this position over the years.

In the first it is asserted that you evidently had a gut instinct leading you to your initial choice, and you should always follow your gut. It is rare for this argument to be offered in jest. Giving credence to ill-defined, quasi-mystical gut instincts seems to be a commonplace error in reasoning for people. My only response is to suggest that for the purposes of concocting a reasonable mathematical model of the situation, "going with your gut" will not enter into our deliberations.

The second argument is really more of an observation: if you switch and lose, you will feel like such a chump that it is better to avoid the potential embarrassment by not switching. Implicit here is the idea that you will feel like less of a chump if you lose by not switching—a dubious proposition, to be sure. Actually, though, there is a more direct refutation of this argument. Our goal is to maximize our chances of winning, not to minimize the chances of feeling like a chump. Surely that is all that needs to be said.

Apparently it is not just *my* students who think this way. Writing in [39], social psychologists Donald Granberg and Thad Brown reported on research undertaken with 114 undergraduate students. The students were asked to ponder a number of versions of the Monty Hall problem and then to comment on their reasoning. Some of the commonly given answers were revealing. In perusing the following quotation, it will help to know that by the "MHD-gnostic condition" the authors mean the classical version of the Monty Hall problem presented at the start of this chapter (version one). By the "MHD-agnostic" condition they mean version two, considered at the start of the next chapter, in which Monty does not know the location of the car but merely chooses a door at random and happens to reveal a goat. The "RRD-agnostic" condition refers to a variation in which there are two cars and one goat. After the player chooses a door, Monty either knowingly opens a car-concealing door (the RRD-gnostic condition) or chooses a door at random and happens to reveal a car (the RRD-agnostic condition). In this context, "RRD" stands for "Russian Roulette Dilemma." I have omitted the references from the following quotation.

> In commenting on their choices, a few subjects showed evidence of counterfactual thinking. For instance, one subject in the MHD-gnostic condition volunteered the thought, "I wouldn't want to pick the other door because if I was wrong I would be more pissed off than if I stayed with the 2nd door and lost." Another subject in the MHD-agnostic condition asserted, "Never change an answer because if you do and you get it wrong it is a much worse feeling." Two subjects in the RRD-agnostic condition reasoned similarly, "It was my first instinctive

choice and if I was wrong, oh well. But if I switched and was wrong it would be that much worse"; "I would really regret it if I switched and lost. It's best to stay with your first choice." Many subjects in all four conditions indicated their impression that the probability is .50 on the final choice, and they subscribed in one way or another to the idea that one's first thought is generally correct. In a later part of this survey, subjects were asked to estimate the percentage of times when they changed their answers on multiple choice tests that the change was from a right to a wrong answer, from a wrong to a right answer, and from a wrong to a wrong answer.... This is despite evidence from research on testing that indicates that changes from wrong to right on multiple choice tests outnumber changes from right to wrong by a ratio of more that 2:1.

These sentiments are indicative of something deep in human psychology. Decision theorists have long noted that people feel much worse if they take an action that leads to negative outcome X than they do if they merely allow X to occur through inaction on their part. In the Monty Hall problem, it is one thing if your initial choice is incorrect and you lose the game by failing to switch. It is quite another to be sitting on the correct door and then lose the game by moving away from it. That is how most people see things, at any rate. For a study of this aspect of the Monty Hall problem, I recommend the paper by Gilovich, Medvec, and Chen [34]. (It shall be discussed in Chapter 6.)

The lesson from this section is that simple, intuitive arguments are not adequate. There is no substitute to a detailed, mathematical consideration of the problem.

2.3. Probability Basics

The time has come to delve into the mathematical issues raised by the Monty Hall problem.

Our understanding of probability begins by envisioning some repeatable experiment with many possible outcomes. We will also assume the number of possible outcomes is finite. Our goal is to make some meaningful statement about how likely certain outcomes are relative to other possibilities. The term "experiment" is to be construed loosely. We do not intend a laboratory situation with beakers and test tubes. Rather, any repeatable activity that can result in many possible outcomes will be considered an experiment for our purposes. Standard examples include rolling dice, tossing coins, or drawing cards out of decks. The Monty Hall problem fits this description. We might imagine playing the game over and over again, always following the same strategy. Our goal is to determine the frequency with which we will win.

That is the setup. Our main question is this: how ought we to assign probabilities to individual events?

It seems natural to say that the probability of getting heads upon one toss of a fair coin is $\frac{1}{2}$, since we have no basis for thinking that one side of the coin is more likely to come up than the other. Likewise, we might say the probability of drawing a diamond from a well-shuffled deck of cards is $\frac{1}{4}$, since there are four suits and each is as likely as any other to be selected.

Generalizing from these examples, it seems reasonable that a probability should be a rational number, that is, a fraction in which both top and bottom are integers. (Among mathematicians, incidentally, it is commonplace to refer to the *top* and *bottom* of a fraction. Jawbreakers like "numerator" and "denominator" figure far more prominently in middle-school mathematics classes than they do in actual mathematical practice.) The bottom of the fraction represents the number of possible outcomes, while the number on top represents the number of outcomes in which the event of interest occurs. In flipping a coin there are two equally likely outcomes, only one of which is heads. In drawing a card from a deck, there are 52 equally likely outcomes, 13 of which are diamonds. This leads to a probability of $\frac{13}{52}$ or $\frac{1}{4}$.

Seen in this way, a probability of 0 could occur only if none of the possible outcomes conforms to our event. Thus, an event has probability 0 only if it is impossible for it to occur. A probability of 1 could occur only if every possible outcome conforms to the event of interest, so the top and bottom of our fraction have the same value. Consequently, an event has probability 1 only if it is certain to occur. And since the number of ways in which our event might occur must be smaller than the total number of possible outcomes of our experiment, we see that a probability must always lie between 0 and 1, inclusive.

Thus, any probability calculation begins by determining both the number of outcomes that are possible from our experiment and the number that conform to our desired event. This is stated succinctly by saying that the probability of an event is the ratio of favorable outcomes to possible outcomes. This is sometimes referred to as the "classical definition of probability," since it is the definition used in many of the historically earliest treatments of this subject.

There are other things to consider, however.

What is the probability of tossing a die and having it come up three? It seems obvious that it is $\frac{1}{6}$, but in answering this way we have simply assumed that we are rolling a standard, six-sided die. The answer would surely change if we were rolling, say, a twelve-sided or twenty-sided die.

So let us assume we are rolling a six-sided die. Can we now conclude the probability of rolling a three is $\frac{1}{6}$? Well, dice can be doctored in various ways to make certain numbers more likely to come up than others. Our assertion of a $\frac{1}{6}$ probability of rolling a three is tacitly based on the assumption that our die has not been so altered. That is, we must assume that each face is as likely to come up as any other face.

These straightforward considerations lead to two conclusions. In measuring the probability of an event, we must first be aware of all the possible outcomes that might happen in lieu of the one in which we are interested. The set of all possible outcomes will be referred to as the **sample space** for the experiment. Determining the probability of obtaining a three on a roll of a die requires prior knowledge of the sample space in which three resides. Is three one of six possible outcomes? Or is it one of twelve or twenty or some other number of outcomes?

This is insufficient, however. We also need some knowledge of the relative likelihoods of the various outcomes. Defining the probability of an event as the ratio of favorable outcomes to possible outcomes makes sense only if the possible outcomes are all equally likely. Recall that we have already seen the perils of ignoring this consideration in our discussion of the Bertrand box paradox. Likewise in our discussion of d'Alembert's proposed probability of obtaining at least one head in two tosses of a coin. In both cases people were led astray by assigning equal probabilities to things that were, in fact, not equally likely.

If knowledge of the probabilities of the outcomes is necessary just to get the ball rolling, then from where is this knowledge supposed to come? There is only one possible answer: it must come from the real-life situation we are trying to understand. If we are rolling a six-sided die that is assumed to be fair, then it seems reasonable to say that each face has the same probability of coming up as any other face. If we know the die is loaded in some way, we must use that information to determine more realistic starting assumptions.

Seems reasonable? Yes, I am sorry, but that is the best we can do. You see, our intent here is to construct an abstract model of a real-world situation. If we are successful, the model will be simultaneously simple enough to comprehend and sufficiently complex to capture the most important aspects of the real-world situation. We can only hope to make our abstract model as realistic as possible based on the information we have. The conclusions we draw from our model will, it is hoped, be testable against actual data collected from practical experimentation. If our theoretical conclusions are consistently wide of the mark, then we will have good reason to go back and modify our model. We will return to this point in later sections.

Now let us suppose that we have defined a sample space and have assigned probabilities in some reasonable way to each of the possible outcomes. The manner in which we assign probabilities to the individual events will be referred to as the **probability distribution** for the space. The combination of a well-defined sample space equipped with a probability distribution will be referred to as a **probability space**.

Sometimes our interest lies not with the probability that a particular outcome will occur, but rather with the probability that at least one of several outcomes will occur. Let us use the term **event** to denote any subset of our sample space. Then we can define the probability of the event to be the sum of the probabilities of the points within that set. Thus, we can now talk about the probability of an event in cases where the event is not merely a simple outcome of the experiment.

In making this definition, however, we have imposed another restriction on the manner in which we assign probabilities to individual outcomes. An event that is certain to happen should have a probability of 1. If we take as our event the set of all points in the space (that is, if we seek the probability that the outcome of our experiment will be something in the sample space), then we should obtain an answer of 1. It is certain, after all, that whatever happens will be something in the space. That means the sum of the probabilities of all of the events in our space must be 1. This is why, for example, we assign a probability of $\frac{1}{n}$ to each outcome in a sample space containing n equally likely outcomes.

2.4. Examples

After all of our theorizing in the previous section, a few examples seem in order.

Let us suppose we flip a fair coin. The sample space would consist of the two possibilities heads and tails, which we will denote by H and T. Since we are assuming the coin is fair, our probability distribution would assign a probability of $\frac{1}{2}$ to each outcome.

Now suppose we flip the coin three times. What is the probability that heads will come up exactly twice?

There are eight possible outcomes of flipping a coin three times:

$$(HHH) \ (HHT) \ (HTH) \ (HTT)$$
$$(THH) \ (THT) \ (TTH) \ (TTT)$$

These eight triples are the elements of our sample space. It seems reasonable to assume that each of these eight possibilities is as likely as any other. Consequently, our probability distribution assigns a probability of $\frac{1}{8}$ to each of these triples. We now seek the probability that the event

$$X = \{(HHT), (HTH), (THH)\}$$

occurs, since these are the triples in which H occurs exactly twice. Since X contains three elements, each with a probability of $\frac{1}{8}$, we conclude that X occurs with probability $\frac{3}{8}$.

Let us consider a more difficult example. Suppose I roll two dice, one of them red and the other one blue. The possible outcomes of this experiment can then be represented by ordered pairs whose first element records the number on the red die, and whose second number records the number on the blue die. That makes thirty-six possible outcomes, which we list below:

$$(1, 1) \ (1, 2) \ (1, 3) \ (1, 4) \ (1, 5) \ (1, 6)$$
$$(2, 1) \ (2, 2) \ (2, 3) \ (2, 4) \ (2, 5) \ (2, 6)$$
$$(3, 1) \ (3, 2) \ (3, 3) \ (3, 4) \ (3, 5) \ (3, 6)$$
$$(4, 1) \ (4, 2) \ (4, 3) \ (4, 4) \ (4, 5) \ (4, 6)$$
$$(5, 1) \ (5, 2) \ (5, 3) \ (5, 4) \ (5, 5) \ (5, 6)$$
$$(6, 1) \ (6, 2) \ (6, 3) \ (6, 4) \ (6, 5) \ (6, 6)$$

Once again we have no basis for assuming that any of these outcomes is more likely than any other, so we assign a probability of $\frac{1}{36}$ to each of them.

Armed with our sample space and distribution, it is now straightforward to answer questions like: What is the probability that the sum of the numbers on the dice is seven? We can survey the thirty-six possibilities above and discover that in six cases we obtain a sum of seven. The probability is therefore $\frac{6}{36}$ or $\frac{1}{6}$. Similarly, we can define X to be the event that a six appears on at least one of the dice. Then the probability of X is determined by noting that X contains 11 elements, each of which has probability $\frac{1}{36}$. Consequently, X has probability $\frac{11}{36}$.

What is the probability that the sum of the numbers on the two dice is prime? Well, the prime numbers between 2 and 12 are $2, 3, 5, 7$, and 11. Examining our thirty-six possibilities above shows that in fifteen cases we obtain a prime number, leading to a probability of $\frac{15}{36}$ or $\frac{5}{12}$.

Let us move on now to playing cards. What is the probability that a randomly chosen card from a standard fifty-two-card deck will be a seven? What is the probability that it will be a face card (that is, a jack, queen, or king)?

All of these questions are answered easily by enumerating the sample space and assuming that each card is as likely to be drawn as any other card. The sample space contains fifty-two elements, one for each card in the deck. Among those fifty-two cards there are 4 sevens and 12 face cards. So the answers to the two questions above are $\frac{4}{52}$ and $\frac{12}{52}$ respectively.

This all seems sufficiently straightforward. Keep in mind, however, that these examples were chosen precisely because of their simplicity. In practical applications it is often quite difficult to choose an appropriate probability distribution. The correct sample space and distribution for the Monty Hall problem are not so readily apparent.

2.5. Solving the Classical Version

How should we model the Monty Hall problem?

In playing the game you first choose a door. Then Monty opens a door, following certain well-specified rules for selecting the door to open. At this point you make your decision about whether or not to switch and your final door choice is opened. Consequently, the possible outcomes of the game can be recorded as ordered triples. Each element of the triple will represent the numerical label on one of the doors. The first element will represent your initial choice. The second element will be the door Monty opens. The third will be the location of the car. That is, each triple in our sample space will have the following form:

(Your initial choice, The door Monty opens, The location of the car).

Since Monty always opens an empty door, we know that the second and third elements in any triple must be different. That is, the door Monty opens cannot also be the door that conceals the car. Likewise, since Monty never opens the door you initially choose, we can assume that the first and second numbers are different as well.

We can now enumerate the possibilities as follows:

$$(1, 2, 1) \ (1, 3, 1) \ (1, 2, 3) \ (1, 3, 2)$$
$$(2, 1, 2) \ (2, 3, 2) \ (2, 1, 3) \ (2, 3, 1)$$
$$(3, 1, 3) \ (3, 2, 3) \ (3, 1, 2) \ (3, 2, 1)$$

Those twelve triples represent our sample space. But how ought we to distribute our probabilities?

The triples in which the first and third entries are different represent scenarios in which you will win by switching doors. They are the cases, after all, where your initial choice is different from the location of the car. Inspecting our list of possibilities shows that there are six such scenarios, and consequently that you will win half the time by switching. This seems to prove that it makes no difference whether or not you switch doors. This assumes, of course, that the twelve possibilities listed above are equally likely. Is that reasonable?

To answer that question, let us simplify our problem by assuming you always choose door number one initially. Making this assumption allows us to confine our attention to the first row of the matrix above. Our sample space now contains a mere four possibilities, specifically

$$(1, 2, 1) \ (1, 3, 1) \ (1, 2, 3) \ (1, 3, 2).$$

To make these scenarios easier to think about, I have also recorded them in Table 2.1.

Now, it is built into the statement of the problem that the car is equally likely to be behind any of the three doors. That means that regardless of your initial choice, the car will be behind door number one exactly one-third of the time. This fact should be reflected in our choice of a probability distribution. Alas, assigning an equal probability to each of the four scenarios would imply that the car is behind door number one precisely half the time. This

Table 2.1: Scenarios in the classical Monty Hall game in which the player initially chooses door one

Your Initial Choice	Monty Opens	Car Is Behind
1	2	1
1	3	1
1	2	3
1	3	2

conclusion is not reasonable, and we are forced to abandon the assumption of equal probabilities.

The trouble arises because correctly choosing door number one leaves Monty with a choice of doors to open. By contrast, in each case where you incorrectly choose door number one, Monty is forced to open the only remaining incorrect door. Still assuming you always choose door one initially, we see there is only one scenario in which the car is actually behind door three, namely, (1, 2, 3). Since the car is behind door three precisely one-third of the time, we should assign a probability of $\frac{1}{3}$ to (1, 2, 3). By similar logic, we should assign a probability of $\frac{1}{3}$ to (1, 3, 2) as well.

That leaves (1, 2, 1) and (1, 3, 1). In both of these cases the car is behind door one, an event that occurs only one-third of the time. Consequently, if X is the event {(1, 2, 1), (1, 3, 1)}, then the probability of X should be $\frac{1}{3}$. It is included in the statement of the problem that in those situations where your initial choice is correct, Monty chooses randomly from the two doors available to him. It follows that the probabilities of (1, 2, 1) and (1, 3, 1) should be equal, and their sum should be $\frac{1}{3}$. Therefore, they should each be assigned a probability of $\frac{1}{6}$.

In stating the problem I have on several occasions stressed the importance of assuming, first, that the doors are initially equiprobable and, second, that Monty chooses randomly when given a choice of doors. Notice that we made explicit use of both of these assumptions in determining the correct probability distribution for our sample space. If either or both of those assumptions are altered, then it is quite possible that a different distribution would be appropriate, and consequently that a different course of action for the player would be indicated.

Since we have a clearly enumerated sample space and an appropriately chosen probability distribution, we can now solve the problem.

What is the probability that you will win by switching, if we assume that you always choose door number one initially? In this case the event of interest is $X = \{(1, 2, 3), (1, 3, 2)\}$, since in the remaining two scenarios you lose by switching. Each of the elements of X has probability $\frac{1}{3}$, implying that X occurs with probability $\frac{2}{3}$. It follows that you double your chances of winning, from one-third to two-thirds, by switching doors. Of course, the assumption that you initially choose door number one is not relevant to that conclusion. Had we started by assuming that you choose door number two or door number three, the result would have been exactly the same.

Therefore, in the classical Monty Hall problem you gain a significant advantage by switching doors when you have the chance.

2.6. Monte Carlo Methods

Perhaps there is a lingering doubt. Granted, the calculations show you gain an advantage by switching doors. But those calculations were carried out on an abstract model, not on the real-world situation. In section 2.2 I even conceded a certain subjectivity in our choice of probability distribution. The intuitive argument, that the two unopened doors are equally likely, remains undaunted. If my abstract calculations imply there is a significant advantage to switching doors, then so much the worse for my calculations. Perhaps I have produced evidence not for the advantage to be gained from switching doors but rather for my lack of skill in choosing probability distributions. How do I resolve this dilemma?

What is needed is actual data against which I can test the predictions of my model. Happily, such data are easily produced. The scenario described by the Monty Hall problem is easily programmed into a computer, and we can run multiple trials in a short amount of time. By keeping track of our wins and losses for different strategies, we can draw some conclusions about the reliability of our model.

In tables 2.2 and 2.3 (pages 50–51) you will find the results of one thousand computer trials of the Monty Hall problem. For the first five hundred I followed a strategy of always initially choosing door number one and sticking with it when given the option of switching. For the next five hundred I always started with door number two and switched when given the chance. In both tables the first column contains the number of trials, the second column contains the number of times I won by following my strategy, and the third contains the winning percentage.

According to my calculations, I should win roughly one-third of the time by following the sticking strategy and roughly two-thirds of the time by following the switching strategy. The results of my simulation confirm these expectations. This is strong evidence that my abstract model is correct in this case.

In simulating the problem I twice needed the services of a random-number generator. First, the car had to be placed randomly behind one of the three doors. Then, in those situations where Monty had a choice of door to open, I had to be certain his choice was made randomly. Methods of this sort, in which multiple trials based in some way on randomness are used to obtain data relevant to the solution of a problem, are referred to as Monte Carlo methods. (We shall leave aside the thorny question of how computers generate "random" numbers.)

Table 2.2: Data from Monty Hall simulation with a uniform sticking strategy

Trials	Wins	Percentage
5	3	.600
10	3	.300
15	5	.333
20	8	.400
25	10	.400
30	12	.400
50	20	.400
70	26	.371
90	31	.344
100	35	.350
120	43	.358
140	49	.350
160	56	.350
180	63	.350
220	75	.341
260	88	.338
300	105	.350
340	117	.344
380	128	.337
420	141	.336
460	156	.339
500	170	.340

By employing such methods I am establishing a connection between the probability of X (which was defined abstractly in terms of a sample space and a probability distribution) and the measured frequency of X in a long run of trials. In other words, if X is an event that might occur in the course of carrying out some repeatable experiment, then I am interpreting the statement "Event X occurs with probability $\frac{a}{b}$" to mean that in a long run of trials, the fraction of the outcomes in which X occurs ought to be very close to $\frac{a}{b}$.

Let us make this more precise. In constructing a probabilistic model, we begin by enumerating a sample space. Each point in the space represents a possible outcome. We next assign a probability distribution to the space. The manner in which we assign the distribution is based in some sensible way on the real-world situation. Once this machinery is in place, we can calculate the probability of X. Let us refer to the probability of X calculated in this way as the "theoretical probability" of X.

Alternatively, we could have carried out multiple runs of the experiment, being careful to keep track of our outcomes. After some large number of trials we could simply count the number of times the event X occurred. In this way we arrive at another value for the probability of X. Let us refer to the ratio of the number of occurrences of X to the number of runs of the experiment as the "actual probability" of X.

The intuition here is that the more trials I carry out, the more likely it is that the theoretical and actual probabilities will match. Over a small number

Table 2.3: Data from Monty Hall simulation with a uniform switching strategy

Trials	Wins	Percentage
5	5	1.00
10	8	.800
15	11	.733
20	14	.700
25	17	.680
30	19	.633
50	33	.660
70	44	.629
90	58	.644
100	65	.650
120	78	.650
140	90	.643
160	105	.656
180	118	.656
220	147	.668
260	169	.650
300	190	.633
340	219	.644
380	242	.637
420	267	.636
460	294	.639
500	321	.642

of trials they may differ greatly, but as the number of trials becomes larger and larger, it becomes almost a sure thing that the two probabilities will coincide.

As an example, think of flipping a coin. The theoretical probability that the coin will land heads is $\frac{1}{2}$. If we flip the coin four times and obtain three heads and one tails, we would compute an actual probability of heads of $\frac{3}{4}$ for this set of trials. Over a mere four tosses it is not surprising if heads comes up three times. But now imagine that we flipped the coin one thousand times. If seven hundred fifty of the tosses landed heads, we would be very suspicious. We expect the number of heads to be close to five hundred in this case. And if we flipped the coin one million times, we would be more surprised still if the observed probability differed significantly from the theoretical probability.

You see this as well in our data from the Monty Hall simulation. In the first five runs employing the strategy of switching doors when given the chance, I won all five times. From this data I would compute an actual probability of 1 and conclude that it is a sure thing that I will win. After ten trials I was still winning 80% of the time. However, it did not take long for the actual probabilities to settle in to values close to the theoretical probability of 66.6%. Indeed, from one hundred trials on the actual probability was always within 5% of the theoretical probability.

This intuition, that the actual probability will eventually come to match the theoretical probability, can be formulated as a mathematical theorem and proved rigorously. The result is then known as the "law of large numbers."

Since proving this would require considerably more machinery than we currently have at our disposal, we will follow the example of the great mathematician Jakob Bernoulli, who described the law of large numbers as so simple as to hardly be worth bothering with. In his book *The Art of Conjecturing* [8] he wrote:

> Neither should it escape anyone that to judge in this way concerning some future event it would not suffice to take one or another experiment, but a great abundance of experiments would be required, given that even the most foolish person, by some instinct of nature, alone and with no previous instruction (which is truly astonishing), has discovered that the more observations of this sort are made, the less danger there will be of error. But although this is naturally known to everyone, the demonstration by which it can be inferred from the principles of the art is hardly known at all, and, accordingly, it is incumbent upon us to expound it here.

I think I will try that line in one of my classes someday. "What do you mean, you don't understand the homework? Even the most foolish person instinctively knows how to do this..."

2.7. Other Approaches

I had two goals in this chapter. One was to solve the Monty Hall problem; the other was to present the basics of probability theory. The method presented here for solving the problem was chosen with those two goals in mind. However, there are other ways of viewing the problem, some of which might seem simpler than what I have presented so far. The professional literature contains a wealth of approaches to this problem, and we shall consider some of them now.

I should also point out that many of the arguments presented below can be found in more than one reference. In attributing them to particular sources I am saying only where I encountered the argument, and am not making any claim about its historical origin.

The ever-useful Wikipedia [98] suggests the following way of enumerating the sample space, based on the idea of treating the two goats as distinct entities:

- The player originally picked the door hiding the car. The game host has shown one of the two goats.

- The player originally picked the door hiding goat A. The game host has shown goat B.

- The player originally picked the door hiding goat B. The game host has shown goat A.

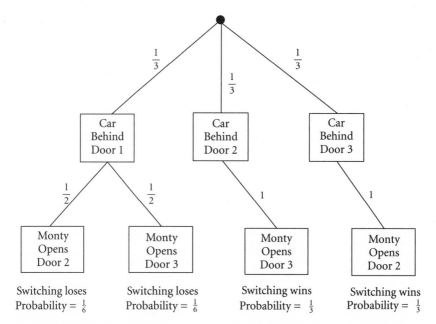

Figure 2.1: Probability tree for the classical Monty Hall problem when the player initially chooses door one

Since the player will select the car, goat *A*, or goat *B* with equal probability, we can treat the scenarios above as equally likely. In the first scenario, the player will lose by switching, while in the remaining two the player wins by switching. Consequently, you win with probability $\frac{2}{3}$ by switching doors.

If you are worried that some skullduggery occurred in the first bullet point, in which two possible scenarios were seemingly combined into one, you might prefer the decision tree diagram in figure 2.1. It provides a useful way of visualizing the probability assignments we made back in section 6.

Also from the Wikipedia article, it ought to be observed that a Monte Carlo simulation for the Monty Hall problem can be performed without a computer. One simply removes three cards from a deck, say the ace of spades and the two red queens. The person in the role of the player selects one of the cards at random and places it to one side. The host, who gets to look at the cards, then discards a red queen from his hand. If he has a choice of queens to discard, he should flip a coin to determine which one goes. If the ace of spades remains in the host's hand, then this corresponds to a time when the player would win by switching. By the law of large numbers, multiple runs of the experiment should lead to a reliable estimate of the probability of winning by switching.

Viewed in this way, whether or not switching results in a win is already determined at the moment after the player selects his card. At that moment the host has either the two red queens in his hand or one red queen and the ace

Table 2.4: Possible locations of the cars and goats

Door 1	Door 2	Door 3
G_1	G_2	C
G_2	G_1	C
G_1	C	G_2
G_2	C	G_1
C	G_1	G_2
C	G_2	G_1

of spades. The former case happens one-third of the time and corresponds to scenarios in which switching leads to a loss. The latter case happens two-thirds of the time and corresponds to cases where switching leads to a win.

The idea of distinguishing between the two goats is the basis for the solution presented by Steven Krantz in [51]. If we identify the contents of the various doors by C, G_1, or G_2 (for car, goat one, and goat two, respectively), then there are six possible arrangements, as shown in Table 2.4:

All six of these layouts are equally likely, and therefore occur with probability $\frac{1}{6}$. We can assume, without loss of generality, that the contestant begins by choosing door three. If we are in the first line of the table, then Monty will now open either door one or door two, and the contestant will lose by switching doors. The situation in the second row is nearly identical, and the player will once again lose by switching.

In the third line Monty will be forced to reveal the goat behind door one. In this case the contestant will win by switching doors. Likewise for the fourth line. The last two lines lead to analogous reasoning, the sole difference being that this time Monty is forced to reveal the goat behind door two.

The final tally is four scenarios in which switching leads to a win against two in which switching leads to a loss. Thus, you win with probability $\frac{2}{3}$ by switching. An elegant way of viewing the problem

2.8. Final Comments

If you are still having trouble seeing the logic behind switching, then perhaps you will be comforted to know that no less an authority than Paul Erdös, undoubtedly one of the greatest mathematicians of the twentieth century, also managed to get it wrong the first time around. The following anecdote is taken from [43]:

Vazsonyi told Erdös about the Monty Hall dilemma. "I told Erdös that the answer was to switch," said Vazsonyi, "and fully expected to move to the next subject. But Erdös, to my surprise, said 'No, that is impossible. It should make no difference.' At this point I was sorry I brought up the problem, because it was my experience that people get excited and emotional about the answer, and I end up with an unpleasant situation. But there was no way to bow out, so I showed him the decision tree solution I used in my undergraduate Quantitative Techniques of Management course." Vaszonyi wrote out a "decision tree," not unlike the table of possible outcomes that vos Savant had written out, but this did not convince him. "It was hopeless," Vaszonyi said. "I told this to Erdös and walked away. An hour later he came back to me really irritated. 'You are not telling me *why* to switch,' he said. 'What is the matter with you?' I said I was sorry, but that I didn't really know why and that the decision tree analysis convinced me. He got even more upset." Vaszonyi had seen this reaction before, in his students, but he hardly expected it from the most prolific mathematician of the twentieth century.

My students would surely sympathize with Erdös' reaction. The book goes on to recount how Erdös became convinced of the correct answer via a Monte Carlo simulation, but that he was still dissatisfied with the result.

Let us close with one further mathematical point. In version one of the Monty Hall problem I explicitly assumed that when Monty has a choice of which door to open he chooses randomly between them. This led us to assign a probability of $\frac{1}{6}$ to each of the outcomes (1,2,1) and (1,3,1) in our sample space. You might have noticed, however, that this assumption was a bit stronger than necessary to justify our conclusion regarding the benefits of switching. Any probability distribution in which the events (1,2,1) and (1,3,1) were assigned probabilities whose sum is $\frac{1}{3}$ would have led to the same conclusion. This point is made clearly in [47].

3

Bayesian Monty

3.1. A Variation

Version Two: As before, Monty shows you three identical doors. One contains a car, the other two contain goats. You choose one of the doors but do
not open it. This time, however, Monty does not know the location of the
car. He randomly chooses one of the two doors different from your selection
and opens it. The door turns out to conceal a goat. He now gives you the
options either of sticking with your original door or switching to the other
one. What should you do?

My obsession with the Monty Hall problem began when I heard this version, and once again I have my father to thank. It seems he had been discussing
the classical version with some friends. Predictably, they offered the standard
argument for why there is no advantage to be gained from switching. The
more my father tried to convince them they were wrong, the more annoyed
they got. Being perfectly familiar with that reaction, I chuckled to myself as I
listened to the story.

That was when my father casually mentioned that if Monty does not
know the location of the car, then it is correct to say there is no advantage to be
gained from switching. I stopped chuckling. In fact, I did a bona fide double
take. What possible relevance could Monty's knowledge make? I figured I
must have heard him wrong.

Well, I hadn't. What Monty knows, or more precisely the presence of a nonzero probability that Monty will open the door concealing the car, makes a big difference to the solution. The purpose of this chapter is to convince you of that fact. Since that will require some serious mathematical machinery, let us first develop some intuition.

As in the classical version, this one begins with a choice from among three identical doors. Your choice will be correct one-third of the time. That means it will be incorrect two-thirds of the time. Those are the cases where, in the classical version, you would win by switching doors.

Let us now assume that your initial choice was incorrect. Then the car is definitely behind one of the remaining two doors. If Monty now happens to choose the door with the car, then the game ends prematurely right there. This will happen in roughly one-half of the cases where you initially make the wrong choice. This makes sense. One-half of two-thirds is one-third, and this represents the fraction of cases in which Monty randomly chooses the door with the car. Since he does not know where the car is, this is a reasonable conclusion.

It also means that among all the trials in which you would have won by switching in the classical version, half are now wasted on trials in which Monty carelessly opens the door with the car. Since in the classical version you win by switching twice as often as you win by sticking, this suggests that in the current version there is no advantage to switching doors.

A plausible agument, but it is no substitute for a mathematically rigorous proof. Our first steps toward such an argument begin with the next section.

3.2. Independent Events

Let us suppose we have two experiments going on at the same time. In one a fair coin is tossed. In the other, a fair six-sided die is rolled. What is the probability of obtaining simultaneously a heads on the coin toss and an even number on the die roll?

As in the previous chapter, we begin our analysis by enumerating the sample space. The possible outcomes of this double experiment can be viewed as ordered pairs in which the first component represents the result of the coin toss and the second component represents the result of the die roll. There will be twelve possibilities, enumerated as follows:

$$(H, 1) \ (H, 2) \ (H, 3) \ (H, 4) \ (H, 5) \ (H, 6)$$
$$(T, 1) \ (T, 2) \ (T, 3) \ (T, 4) \ (T, 5) \ (T, 6)$$

It seems reasonable to assume that the result of the coin toss and the result of the die roll do not affect each other. That is, we do not have a situation in which knowing the coin landed heads makes it seem more likely that we got an even number on the die roll or anything like that. The appropriate probability distribution is therefore the uniform one, which in this case means every outcome is assigned a probability of $\frac{1}{12}$. Events of this sort, where the occurrence of one has no bearing on the occurrence of the other, are said to be independent.

We can now proceed as in the first chapter. Among the twelve equally likely ordered pairs there are three with an H in the first component and an even number in the second. Consequently the probability of obtaining both a head and an even number is $\frac{3}{12}$ or $\frac{1}{4}$.

You have probably noticed something interesting, however. The coin toss has two possible outcomes. The die roll has six possible outcomes. The two experiments performed simultaneously therefore have twelve possible outcomes. This is a special case of a general counting rule with which you are probably familiar: If a finite set A has x elements and a finite set B has y elements, then the number of ordered pairs whose first element comes from A and whose second element comes from B is given by xy. We will use the notation $A \times B$ to denote this set of ordered pairs. We will also speak of the **Cartesian product** of the sets A and B.

Now, the probability of obtaining heads on a coin toss is $\frac{1}{2}$ and the probability of obtaining an even number on a single roll of a die is likewise $\frac{1}{2}$. Multiplying these together gives $\frac{1}{4}$, which is precisely the value we just computed. This logic applies more generally.

When we say that event E_1 occurs with probability $\frac{a}{b}$, we imagine an experiment with b possible outcomes, a of which correspond to an occurrence of E_1. Likewise, if E_2 occurs with probability $\frac{c}{d}$, we imagine d possible outcomes, with c of them corresponding to an occurrence of E_2.

If these experiments take place independently of each other, then bd will represent the total number of possible outcomes when the two experiments are performed simultaneously. And since E_1 can occur in a different ways, while E_2 can occur in c different ways, we obtain a probability of $\frac{ac}{bd}$ for the simultaneous occurrence of E_1 and E_2.

Keep in mind that events are defined to be subsets of the sample space. Therefore, the intersection $E_1 \cap E_2$ of two events represents the event in which both E_1 and E_2 occur simultaneously. For example, in the experiment of tossing a coin and rolling a die, we might view E_1 as the event of getting a heads and E_2 as the event of rolling an even number. Then we would have

$$E_1 = \{(H, 1), (H, 2), (H, 3), (H, 4), (H, 5), (H, 6)\}$$
$$E_2 = \{(H, 2), (T, 2), (H, 4), (T, 4), (H, 6), (T, 6)\}.$$

Consequently,

$$E_1 \cap E_2 = \{(H, 2), (H, 4), (H, 6)\}$$

corresponds to the event in which both E_1 and E_2 occur.

Intuitively, two events are independent if they are not related to each other. Alas, it is not always so easy to determine whether two events meet this criterion. What is needed is a more mathematically precise notion of independence. And since we have just seen that the probability that two independent events happen simultaneously is simply the product of the individual probabilities, we may as well use *that* as our definition.

Definition 1   Let A and B be two events in a probability space. We say that A and B are **independent** if

$$P(A \cap B) = P(A)P(B).$$

Natural though it is, there is something a bit irritating about this. The determination that two events are independent depends, according to this definition, on the probability distribution we have chosen. The events A and B can be independent with respect to one distribution but not independent with respect to some other distribution.

Does this not strike you as odd? The knowledge of which events are independent of one another ought to affect our choice of a probability distribution. Yet our definition makes it impossible to determine that two events are independent until *after* a distribution has been chosen. We seem to have done things backward.

The resolution to this dilemma lies in recognizing the difference between an abstract model and the reality it seeks to describe. We have an intuitive notion of independence, in which events are independent if they have no connection to one another. This notion is very useful in selecting an appropriate distribution in the first place. But sometimes it is not so obvious whether two abstractly defined events are independent. In that case, so long as we have a decent grasp on the probabilities with which our simple events occur, we can use the mathematical definition to ferret out connections our intuition might have overlooked.

That is always how it goes. We use our understanding of the real-world situation to construct the mathematical model. Then we study the model to better understand the real-world situation. Theory and practice must work hand in hand.

What are some examples of non-independent events?

Suppose we choose a card from a deck. Let A denote the event that the chosen card is red, and let B denote the event that the chosen card is a diamond. These events are not independent. The probability that the chosen card is a diamond goes up when we learn that the chosen card was red.

As another example, let A denote the event that a randomly chosen American citizen is under five feet tall and let B denote the event that a randomly chosen American is under twelve years old. Since the probability that a person is under five feet tall goes up when we learn that he is under twelve years old, these events are not independent.

3.3. Examples

Given the importance of the ideas in the previous section, we should pause to consider some specifics.

Consider again the example of tossing a coin and rolling a die. Let A be the event of tossing a tails and let B be the event of rolling a perfect square. Then $P(A) = \frac{1}{2}$ and $P(B) = \frac{1}{3}$ (since one and four are the only perfect squares on a six-sided die). An inspection of the sample space reveals that

$$P(A \cap B) = P(\{(T, 1), (T, 4)\}) = \frac{2}{12} = \frac{1}{6}.$$

Since $P(A)P(B) = P(A \cap B)$, we have that A and B are independent events, precisely as we would expect.

Let us return now to the classical Monty Hall problem. In the last chapter we enumerated the sample space as follows:

$$(1, 2, 1) \ (1, 3, 1) \ (1, 2, 3) \ (1, 3, 2)$$
$$(2, 1, 2) \ (2, 3, 2) \ (2, 1, 3) \ (2, 3, 1).$$
$$(3, 1, 3) \ (3, 2, 3) \ (3, 1, 2) \ (3, 2, 1)$$

Recall that each triple has the form

(*Your initial choice, The door Monty opens, The location of the car*).

We then simplified our work by restricting our attention entirely to the first row of this matrix. We simply assumed that you always initially choose door number one. By doing so we avoided having to define a probability distribution for the entire sample space. We remedy that now.

Since we are given that the car is placed randomly behind the three doors, we can use the location of the car to partition our triples into three sets as follows:

$$A = \{(1, 2, 1), (1, 3, 1), (2, 3, 1), (3, 2, 1)\}$$
$$B = \{(1, 3, 2), (2, 1, 2), (2, 3, 2), (3, 1, 2)\}$$
$$C = \{(1, 2, 3), (2, 1, 3), (3, 1, 3), (3, 2, 3)\}.$$

Thus A, B, and C represent the events that the car is behind door one, two, or three respectively. Our probability distribution should be chosen so that

$$P(A) = P(B) = P(C) = \frac{1}{3}.$$

We have two further pieces of information with which to work:

1. Your initial choice of door is made independently of the location of the car.

2. When Monty has a choice of doors to open, he chooses randomly from among his choices.

How do we put all this together?

As in Chapter 1, let us focus on the four triples in which you initially choose door one. By item one, the location of the car is independent of this choice. That means that among all of the instances where you initially choose door one, the car is equally likely to be behind any of the three doors. Therefore:

$$P(1, 3, 2) = P(1, 2, 3) = P(\{(1, 2, 1), (1, 3, 1)\}).$$

Item two gives us the further piece of information that

$$P(1, 2, 1) = P(1, 3, 1).$$

Since $P(A) = \frac{1}{3}$, we quickly determine that we must have

$$P(1, 2, 1) = P(1, 3, 1) = \frac{1}{18}, \quad P(1, 3, 2) = \frac{1}{9}, \quad P(1, 2, 3) = \frac{1}{9}.$$

The symmetry of the situation now allows us to fill in the remaining probabilities as follows:

$$P(2, 3, 1) = P(2, 1, 3) = P(3, 2, 1) = P(3, 1, 2) = \frac{1}{9}$$

$$P(2, 1, 2) = P(2, 3, 2) = P(3, 1, 3) = P(3, 2, 3) = \frac{1}{18}$$

Let us investigate some consequences of this distribution. Let S be the event that you initially choose door number one and let T be the event that Monty opens door number two. Since Monty's choice of door depends both on the location of the car and on the door you chose, we would not expect these events to be independent. Our probability distribution confirms this expectation. Note that S corresponds to the event

$$S = \{(1, 3, 2), (1, 2, 3), (1, 2, 1), (1, 3, 1)\}$$

while T is the event

$$T = \{(1, 2, 1), (3, 2, 1), (1, 2, 3), (3, 2, 3)\}$$

Finally, we see that

$$S \cap T = \{(1, 2, 3), (1, 2, 1)\}.$$

Since our distribution tells us the probability assigned to each triple individually, we can readily compute that

$$P(S) = \frac{1}{3}, \quad P(T) = \frac{1}{3}, \quad P(S \cap T) = \frac{1}{6}.$$

In this case we have $P(S)P(T) \neq P(S \cap T)$. Consequently, S and T are not independent.

Now let B denote, as before, the event that the car is behind door number two. Then we have

$$S \cap B = \{(1, 2, 3)\}$$

and

$$P(B) = \frac{1}{3}, \quad P(S \cap B) = \frac{1}{9}.$$

Since we still have $P(S) = \frac{1}{3}$, we see that in this case $P(S)P(B) = P(S \cap B)$, implying that S and B are independent events. Again, this agrees with our intuition. Your choice of initial door is independent of the location of the car.

3.4. Conditional Probability

Currently we have two notions of independence. The first is intuitive: two events are independent if they are not connected to each other in any way. The second is mathematical: two events are independent if the probability that both occur is equal to the products of the probabilities that either one occurs alone.

There is a third way of thinking about independent events: two events are independent if the knowledge that one of them has occurred has no effect on our estimate of the probability that the other has also occurred (or, perhaps, will occur). When we are simultaneously tossing a coin and rolling a die, for example, the knowledge that the coin landed heads will not affect our estimate of the probability that the die came up three.

Let us now revisit the classical Monty Hall problem. For concreteness, suppose we initially choose door number one and Monty then opens door number two. Our initial choice is correct with probability $\frac{1}{3}$. Does the knowledge that Monty has shown that door number two is empty give us any reason to alter our probability assessment? Our analysis in the previous chapter shows that it does not.

But if our intuition is correct, then in version two the same piece of information does give us reason to alter our probability assessments. In both cases we intially choose a door with a $\frac{1}{3}$ probability of being correct. In both cases Monty opens a door and shows us that it is empty. Yet in the first case we do not alter our initial judgment, while in the second case we do. Why is that?

The reason has to lie in the different probabilities attached to Monty's actions. In version one, it was certain that Monty would open an empty door. Not so in version two. Monty now chooses his door randomly, and therefore there is only a $\frac{2}{3}$ probability that he will open an empty door. Unless our intuition has failed us, the difference between the two versions must lie here. But how ought we incorporate this observation into our reasoning?

This is a very general sort of situation. We are interested in some event A. We know that some other event, B, has already occurred. We seek the probability not of A occurring by itself, but rather the probability that A will occur when we know that B has occurred. This is referred to as the **conditional probability** that A will occur, given that B has occurred. More concisely, we refer to "the probability of A, given B." We will use the notation $P(A|B)$ to denote this probability. What is needed is a formula for $P(A|B)$.

Such a formula appears readily if we think in terms of relative frequencies. We are assuming that A and B are events in the sample space of an experiment. Imagine that we have carried out some large number n of trials. Further assume that event B occurred x times. If n is sufficiently large, then the law of large numbers tells us that $P(B)$ is very close to $\frac{x}{n}$.

We now examine those x trials and determine the number of occurrences of A among them. Let us call this number y. In these y trials we have that A and B both occurred. Since this is out of a total of n trials, we can assume that $P(A \cap B) = \frac{y}{n}$.

How should we interpret $P(A|B)$, given our setup so far? We have x trials in which event B has occurred. Among those trials there are y in which event A has occurred. So the relative frequency of A occurrences among those trials in which B has occurred, is given by $\frac{y}{x}$. Consequently, we ought to have $P(A|B) = \frac{y}{x}$.

Of course, these are actual probabilities, meaning that they were determined from data generated by multiple trials of a repeatable experiment. The numbers x and y are variables whose value depends on the specific experiment and the number of trials we have done. That notwithstanding, we can use this reasoning to formulate a reasonable definition for the theoretical value of $P(A|B)$.

Definition 2   Let A and B be events in a probability space. Then we define the **conditional probability** of A given B, denoted by $P(A|B)$, by the formula

$$P(A|B) = \frac{P(A \cap B)}{P(B)}.$$

This definition is certainly consistent with our argument, since we have

$$\frac{P(A \cap B)}{P(B)} = \frac{y/n}{x/n} = \frac{y}{x}.$$

What happens if $P(B) = 0$? Our formula would then require dividing by zero, which is not a good thing. It follows that $P(A|B)$ is undefined when $P(B) = 0$. This is logical, for in that case we would be seeking the probability that A occurs under the assumption that something impossible happens first. On the other hand, what happens if $P(A \cap B) = 0$? Then our formula assigns a probability of 0 to $P(A|B)$. Again, this makes sense. If it is impossible that both A and B occur simultaneously, then the knowledge that B has occurred tells us immediately that A has not occurred.

And what happens if A and B are independent events? Intuitively, this ought to mean that $P(A|B) = P(A)$. Happily, our formula bears that out. For if A and B are independent, then $P(A \cap B) = P(A)P(B)$, and we have

$$P(A|B) = \frac{P(A \cap B)}{P(B)} = \frac{P(A)P(B)}{P(B)} = P(A),$$

precisely as expected. As a bonus, we also have the reverse implication. If $P(A|B) = P(A)$ then we must have that A and B are independent. You might enjoy the challenge of proving that for yourself.

3.5. The Law of Total Probability

The next section will bear witness to the power of conditional probability. First, however, we must add an additional computational weapon to our arsenal.

For motivation we return once more to the classical Monty Hall problem. Let us imagine that you have initially selected door one. What is the probability that Monty will now open door two? The answer is not so clear. Certainly if the car is behind door two there is a probability of zero that Monty will open it. If the car is behind door three, then since you have already selected door one, it is certain that Monty will open door two. And if the car is behind door one? In this case Monty chooses randomly between doors two and three, giving a probability of $\frac{1}{2}$ that Monty opens door two.

The car can be behind any of the three doors, and the probability that Monty opens door two is different in each of the three cases. How is this information to be assembled into a single probability?

Our answer begins by defining a few events. Let C_1, C_2, and C_3 denote respectively the events that the car is behind door one, two, or three. These possibilities exhaust all of the possible locations of the car, and they are mutually exclusive. Denote by M_2 the event in which Monty opens door two. We can then write

$$P(M_2) = P(M_2 \cap C_1) + P(M_2 \cap C_2) + P(M_2 \cap C_3).$$

This follows from the observation that the events $M_2 \cap C_i$ for $i = 1, 2, 3$ are mutually exclusive and collectively exhaustive. That is, if M_2 occurs, it must occur in conjunction with exactly one of C_1, C_2, or C_3.

The definition of conditional probability allows us to rewrite that expression without recourse to any intersection symbols. We obtain

$$P(M_2) = P(M_2|C_1)P(C_1) + P(M_2|C_2)P(C_2) + P(M_3|C_3)P(C_3).$$

Note, now, a remarkable fact. Every term in that sum can be evaluated. Since the doors are equally likely to conceal the car, we have

$$P(C_1) = P(C_2) = P(C_3) = \frac{1}{3}.$$

Our previous reasoning shows that

$$P(M_2|C_1) = \frac{1}{2}, \quad P(M_2|C_2) = 0, \quad P(M_2|C_3) = 1.$$

When everything is plugged into the appropriate slot we obtain

$$P(M_2) = \left(\frac{1}{2}\right)\left(\frac{1}{3}\right) + (0)\left(\frac{1}{3}\right) + (1)\left(\frac{1}{3}\right) = \frac{1}{2}.$$

There is a pleasing logic to this. It says that if we play the game a large number of times and initially choose door one each time, then we expect to see Monty open door two roughly half the time.

The strategy employed here for determining $P(M_2)$ is frequently useful in problems involving conditional probability. In recognition of that fact, we will pause to state it more formally.

Let us suppose that Ω represents the sample space for some experiment. Further suppose that $A_1, A_2, \ldots, A_n$ represent a collection of events that partitions Ω. In other words, they are mutually exclusive and collectively exhaustive. Also assume that all of them represent possible events, so $P(A_i) \neq 0$ for any value of i between 1 and n. Finally, let B be an event satisfying $P(B) \neq 0$. Then we have

$$P(B) = P(B|A_1)P(A_1) + P(B|A_2)P(A_2) + \ldots + P(B|A_n)P(A_n).$$

This is known as the **law of total probability**. Since there is a more general, and more complicated, theorem that also goes by that name, some references refer to this as the "law of alternatives." Whatever you may wish to call it, we will use it frequently in the pages to come.

3.6. More Examples

Suppose you flip two coins and the first one lands heads. What is the probability that the second coin lands heads as well?

Let us define B as the event that the first coin comes up heads and A as the event that the second coin comes up heads. The question asks for $P(A|B)$. We begin by enumerating the sample space. There are only four possibilities:

$$(H, H) \quad (H, T) \quad (T, H) \quad (T, T).$$

If we assume the coin is fair, then each of these events occurs with probability $\frac{1}{4}$.

If we are given that the first coin landed heads, then we have effectively reduced the sample space to just the two events:

$$(H, H) \quad (H, T).$$

Since these events are equally likely, we conclude that $P(A|B) = \frac{1}{2}$. This matches our expectations. The result of one fair coin toss should not affect the result of any other fair coin toss.

Our formula for $P(A|B)$ produces the same answer. We have $P(B) = \frac{1}{2}$ and $P(A \cap B) = \frac{1}{4}$. Therefore, we compute

$$P(A|B) = \frac{P(A \cap B)}{P(B)} = \frac{1/4}{1/2} = \frac{1}{2},$$

as expected.

Now suppose that, once again, you flip two coins. This time, however, you look away before they land. You ask me, "Did at least one of the coins land heads?" I truthfully answer, "Yes." Again we ask for the probability that both coins are heads.

The difference between this example and the last is subtle but important. You see, there are a number of questions we might ask regarding the results of two coin tosses. Let us assume the coins are so labeled so that we may speak of "the first coin" and "the second coin." Then we might ask, "What was the result of the toss of the first coin?" Alternatively, we might ask, "What was the result of the second coin toss?" Knowing the answer to either one of these questions tells us nothing about the answer to the other. That is the situation we confronted in the first problem.

A third question we might ask is, "What is the distribution of heads and tails among the two coin tosses?" No longer are we viewing the coins as individual entities. Rather, we seek information regarding the two coins viewed as an ensemble. The second problem does not ask, "What can you tell me about the outcome of the second coin toss given that the first toss came up heads?" Instead it asks, "What can you tell me about the distribution of heads and tails among the two tosses given that one of the coins came up heads?" These are very different questions indeed.

So, let A once more be the event that both coins land heads and let B be the event that one of the coins landed heads. Learning that B has occurred reduces our sample space to just three possibilities:

$$(H, H) \quad (H, T) \quad (T, H).$$

We are still justified in thinking these three outcomes to be equally likely. That is, when I answer your question by telling you there is indeed at least one head among the two tosses, you are justified in thinking that the outcome TT has not occurred, but you are not justified in thinking one of the three remaining outcomes is more likely than the other two. Consequently, since only one of the three equiprobable outcomes involves two heads, we see that $P(A|B) = \frac{1}{3}$.

Alternatively, we might have used our formula for conditional probability as follows. We have $P(B) = \frac{3}{4}$, since the two coins can land in four equiprobable ways, three of which involve at least one head. Also, since $A \cap B = A$, we have $P(A \cap B) = P(A) = \frac{1}{4}$. So

$$\frac{P(A \cap B)}{P(B)} = \frac{1/4}{3/4} = \frac{1}{3},$$

and once again our formula gives the expected answer.

Let us go one more round. As before, you flip two coins but look away before they land. This time, however, I roll a fair six-sided die. I resolve that if the die comes up even, I will look at the first coin and tell you what I see. If it comes up odd, I will instead tell you about the second coin. You, who know that I am following this procedure, hear me say, "I see a heads." Once again, what is the probability that the other coin also came up heads?

If you are inclined to think this problem is identical to the one just discussed, then I urge you to reconsider. The analysis certainly begins in the same way. The revelation that one of the coins came up heads still leaves us with the three possibilities

$$(H, H) \quad (H, T) \quad (T, H).$$

The difficulty comes in the next step. In the second version of the problem, my statement that there is at least one head had the same probability of being made regardless of which of the three scenarios had occurred. We would have answered "Yes!" regardless of which of the three outcomes had happened. Not so in our new version. If we are in the (H, H) scenario, then we will state that we see a heads regardless of the outcome of the die roll. But if we are in the (H, T) or (T, H) scenario, it is only fifty-fifty that we will report that we see a heads. In light of this we can no longer regard the three scenarios as equally likely.

This presents a bit of a problem. If they are not equally likely, then what probabilities ought to be assigned? We might argue that since the statement "I see a heads," is twice as likely to be made in the (H, H) scenario than it is in either of the other two scenarios. I should assign (H, H) a probability that is twice what I assign to (H, T) or (T, H). That is, I should now think $P(H, H) = \frac{1}{2}$ and $P(H, T) = P(T, H) = \frac{1}{4}$. This argument is correct, as we shall see after we have added Bayes' theorem to our repertoire (see section 3.12).

Our problem may also be resolved via the law of total probability. Proceed as before by defining A to be the event that both coins come up heads and defining B to be the event that I tell you that I see a heads after following the aforementioned procedure. Again we seek $P(A|B)$, which requires determining $P(A \cap B)$ and $P(B)$.

Since B occurs with certainty whenever A occurs, we can say that

$$P(A \cap B) = P(A) = \frac{1}{4}.$$

We evaluate $P(B)$ via the law of total probability, thereby obtaining

$$P(B) = P(B|(H, H))P(H, H)$$
$$+ P(B|(H, T))P(H, T) + P(B|(T, H))P(T, H)$$
$$= (1)\left(\frac{1}{4}\right) + \left(\frac{1}{2}\right)\left(\frac{1}{4}\right) + \left(\frac{1}{2}\right)\left(\frac{1}{4}\right) = \frac{1}{2},$$

which confirms our earlier, intuitive argument. A fascinating result!

This last problem has a clear relationship with the Bertrand box paradox from Chapter 1. Recall that we were there confronted with three boxes, one containing two gold coins, one containing two silver coins, and one containing one gold and one silver. A box is chosen at random, and from that box a random coin is drawn. That coin is seen to be gold. What is the probability that the other coin is also gold? Let us apply our newfound wisdom to *that* little bagatelle.

The key to solving the problem was the observation that you are twice as likely to remove a gold coin from the box with two gold coins than you are to remove it from the box with one gold and one silver. Let us denote by B_{gg} and B_{gs} respectively the events that we have reached into the gold/gold and gold/silver boxes. (It is evident that we have not reached into the silver/silver box.) Let G denote the event of reaching into our chosen box and randomly removing a gold coin from it. Note that we are certain to remove a gold coin if we have chosen the gold/gold box. We can write

$$P(B_{gg}|G) = \frac{P(B_{gg} \cap G)}{P(G)}$$

$$= \frac{P(B_{gg})}{P(G|B_{gg})P(B_{gg}) + P(G|B_{gs})P(B_{gs})}$$

$$= \frac{1/3}{(1)\left(\frac{1}{3}\right) + \left(\frac{1}{2}\right)\left(\frac{1}{3}\right)} = \frac{2}{3},$$

just as we computed in Chapter 1. A most satisfying conclusion.

3.7. Bayes' Theorem

Useful though they are, the definition of conditional probability and the law of total probability do not exhaust our resources. We have not yet ferreted out the connection between $P(A|B)$ and $P(B|A)$, you see.

Let A and B be two events in a probability space. From the previous section we know that

$$P(A \cap B) = P(B)P(A|B).$$

On the other hand, we could just as easily write

$$P(B \cap A) = P(A)P(B|A).$$

If we set the right-hand sides of these equations equal to each other and then divide by $P(B)$, we obtain the formula

$$P(A|B) = \frac{P(A)P(B|A)}{P(B)}. \tag{3.1}$$

This result is known as Bayes' theorem, after the eighteenth-century British mathematician who first formulated a version of it.

Despite its modest appearance, Bayes' theorem is the key to solving a host of probabilistic problems, including version two of the Monty Hall scenario. Look at the left-hand side of equation (3.1). It refers to the probability of A given B. Now look at the right-hand side. The only conditional probability is the probability of B given A. Thus, Bayes' theorem establishes a highly counterintuitive symmetry between the two conditional probabilities. Specifically, it says that you can replace $P(A|B)$ by $P(B|A)$, so long as you multiply by the ratio of the individual probabilities taken in their original order (that is, if you multiply by the fraction $\frac{P(A)}{P(B)}$).

Why should there be a connection between $P(B|A)$ and $P(A|B)$?

Look again at equation (3.1). On the right-hand side we see the term $P(A)$. On the left-hand side we have $P(A|B)$. It is common to refer to $P(A)$ as the **prior probability** of A. It is the probability we assign to A without any knowledge of other events that may have occurred. By contrast, $P(A|B)$ can be viewed as the **updated or posterior probability** we assign to A given the knowledge that B has occurred.

According to (3.1), the updated probability of A is obtained by multiplying its prior probability by the fraction $\frac{P(B|A)}{P(B)}$. This fraction is typically referred to as the **likelihood** of A given B. When that fraction exceeds 1, that is, when $P(B|A) > P(B)$, the updated probability of A is greater than its prior probability. This makes sense. For suppose we have a situation in which knowledge that A has occurred increases our estimation of the probability that B has occurred. We can think of this as establishing a strong correlation between occurrences of B and occurrences of A. Specifically, B occurrences are strongly correlated with A occurrences. In such a situation, would not the knowledge that B has occurred cause us to increase our estimation of the probability that A has occurred as well?

The counterintuitive nature of Bayes' theorem goes away by considering a few practical examples. The probability that a suspect is guilty increases when his fingerprints are found on the gun. Likewise, the probability that a suspect's fingerprints are found on the gun goes up when we learn that he is guilty of the crime. The probability that a person has lung cancer goes up upon learning that he is a cigarette smoker, and the probability that he is a cigarette smoker goes up upon learning that he has lung cancer. In the other direction, the probability that someone suffers from a large number of dental cavities goes down when you learn that he rarely eats sugar. And, as expected, the probability that someone eats a lot of sugar goes down upon learning that he does not suffer from cavities.

Perhaps a more practical example is called for. Imagine you are standing at the baggage retrieval carousel at a major American airport. Let x denote the percentage of bags that have already emerged from wherever they are before appearing in the terminal. How large should x be before the probability is greater than $\frac{1}{2}$ that your bag is not there?

If you are like me, you start worrying very early indeed. While I did once have the satisfaction of seeing my bag emerge first after a long and wearying cross-country flight, I find it is far more common to have my bag arrive fairly late in the game. I travel by air a couple of times each year, and so far I have been fortunate never to have had one of my bags go astray. That notwithstanding, by the time the first bags have completed one revolution around the carousel, a sinking feeling develops in the pit of my stomach. This, as I will show, is not really rational.

Since the total number of bags is not relevant to our calculation, we will simplify matters by assuming there are one hundred bags on the flight. Imagine numbering them from one to one hundred based on the order in which they appear on the carousel (so bag number one is the first bag to appear, and bag number one hundred is the last bag to appear). We will assume that your bag is as likely to receive any of the numbers from one to one hundred as any other. Thus, the probability that your bag is among the first fifty is $\frac{1}{2}$, that it is among the first twenty is $\frac{1}{5}$, and so on.

We define event A to be "Your bag has been misplaced by the airline" and define B to be "Your bag is not among the first x bags to appear on the carousel." We seek the smallest value of x that makes $P(A|B) \geq \frac{1}{2}$. Bayes' theorem tells us that determining $P(A|B)$ requires an evaluation of $P(A)$, $P(B|A)$ and $P(B)$.

Now, it is clear that $P(B|A) = 1$ regardless of the value of x, since if your bag has been misplaced by the airline, it is certain that it will not be among the first x bags.

To evaluate $P(A)$ I consulted a 2006 report from the U.S. Department of Transportation. It seems that the least reliable United States airlines misplace roughly twenty out of every one thousand bags, a rate of 2%.

That leaves only $P(B)$. Since the value of $P(B)$ differs depending on whether or not the airline has lost your bag, we will need the law of total probability to evaluate it. Suppose that the airline has not misplaced your bag. In that case, given our assumption that your bag is as likely to occupy any given place in line as any other, we can think of $\frac{x}{100}$ as representing the probability that your bag is among the first x bags to appear on the carousel.

Consequently, the probability that your bag is not among the first x, assuming that your bag has not been misplaced, is given by $1 - \frac{x}{100}$.

The law of total probability now tells us that

$$P(B) = P(A)P(B|A) + P(\bar{A})P(B|\bar{A})$$

$$= \left(\frac{2}{100}\right)(1) + \left(\frac{98}{100}\right)\left(1 - \frac{x}{100}\right) = 1 - \frac{98x}{10,000}.$$

Plugging everything into Bayes' theorem reveals that

$$P(A|B) = \frac{P(A)P(B|A)}{P(B)} = \frac{\left(\frac{2}{100}\right)(1)}{1 - \frac{98x}{10,000}} = \frac{200}{10,000 - 98x}.$$

There is a certain plausibility to this formula. Our model of the situation implies that we have $0 \leq x \leq 100$. The case $x = 100$ represents the state of affairs after all of the bags have been delivered. The case $x = 0$ is the initial state, before any bags have appeared. If we plug $x = 0$ into our formula, we obtain $P(A|B) = \frac{1}{50} = 2\%$, which is the base rate at which luggage is lost by the airline. On the other hand, plugging in $x = 100$ leads to $P(A|B) = 1$. Again, this is quite right. If all of the bags have appeared and yours is not among them, then it is certain that the airline has lost your bag.

It is now readily computed that the smallest integer value of x for which $P(A|B)$ is smaller than $\frac{1}{2}$ is $x = 98$. That means that even if 97% of the bags have already appeared without yours being among them, the probability that yours is among the final 3% is still greater than $\frac{1}{2}$.

And this, mind you, is the figure assuming the airline loses 2% of the bags that it handles. That is the worst rate among major American airlines. The industry average is closer to .63%. Basically, you should not conclude your bag has been lost until every single bag has been deposited on the carousel and yours is still not there.

We have seen this sort of thing before. In Chapter 1 we considered the problem of a medical test for a disease that afflicts one of every one thousand people in the population. We stipulated that the test returns no false negatives but does return a false positive in 5% of all cases. It turned out that a positive test result implied only a 2% chance of actually having the disease. The base rate at which the disease occurs in the population is so small that a positive test result is far more likely to be the result of an error than of actual sickness.

That is precisely the situation here. The airlines lose such a small percentage of the bags they handle that your bag not appearing early on in the baggage retrieval process is far more likely to be the result of bad luck than of carelessness on the part of the airline.

A final thought: how do you determine the fraction of bags that have already emerged? It can be estimated from the number of agitated people milling around the carousel, of course.

3.8. Monty Meets Bayes

How can we apply Bayes' theorem to the Monty Hall problem?

As a warm-up exercise, consider the state of affairs before you actually sit down to play the game. Let A denote the event "Your initial choice conceals the car," and let B denote the event "Monty opens a door to reveal a goat after you make your initial choice." In version one, we know that Monty is guaranteed to reveal a goat (and chooses his door randomly when he has a choice) and that each door initially has the same probability of concealing the car. In light of that, Bayes' theorem tells us that

$$P(A|B) = \frac{P(A)P(B|A)}{P(B)} = \frac{\frac{1}{3}(1)}{1} = \frac{1}{3},$$

precisely as we found previously.

On the other hand, in version two it is no longer a sure thing that event B will occur. This will affect our evaluation of $P(B)$. That said, we still have $P(B|A) = 1$, since we are certain that Monty will not open our initial choice.

Since the probability that Monty will reveal a goat changes depending on the location of the car, we will need the law of total probability to evaluate $P(B)$. Let $\bar{A}$ denote the event "Your initial choice conceals a goat." Then we have

$$P(B) = P(A)P(B|A) + P(\bar{A})P(B|\bar{A})$$
$$= \frac{1}{3}(1) + \frac{2}{3}\left(\frac{1}{2}\right) = \frac{2}{3}.$$

This time Bayes' theorem gives us

$$P(A|B) = \frac{\frac{1}{3}(1)}{\frac{2}{3}} = \frac{1}{2}.$$

Just as we expected. In version one we double our chances of winning, while in the second version our two remaining doors are equally likely.

Still, perhaps there is a remaining doubt. When we are actually playing the game we have more information at our disposal than when merely thinking about it from our armchairs. For if I am actually playing I know both the door I initially chose and the door Monty opened. Perhaps these details are relevant to our calculations.

In the interests of putting these fears to rest, and with the spirit that our nascent Bayesian muscles need a bit more flexing, let us embark on a more detailed analysis.

We begin by defining, as before, C_1, C_2, and C_3 to be the events in which the car is placed behind doors one, two, or three respectively. Similarly, define M_1, M_2, and M_3 to be the events in which Monty opens doors one, two, or three respectively to reveal a goat after you make your initial choice. For concreteness, let us assume that you initially choose door one and Monty then opens door two. Then we must evaluate $P(C_3|M_2)$. (We could also evaluate $P(C_1|M_2)$, of course, but the calculations are slightly simpler this way.)

Combining Bayes' theorem with the law of total probability gives us

$$P(C_3|M_2) = \frac{P(C_3)P(M_2|C_3)}{P(M_2)}$$

$$= \frac{P(C_3)P(M_2|C_3)}{P(C_1)P(M_2|C_1) + P(C_2)P(M_2|C_2) + P(C_3)P(M_2|C_3)}.$$

If we assume the three doors are equally likely, then we have

$$P(C_1) = P(C_2) = P(C_3) = \frac{1}{3}.$$

If we further assume that Monty will open neither the door with the car nor the door you initially chose, then we have

$$P(M_2|C_2) = 0 \text{ and } P(M_2|C_3) = 1.$$

If we plug these values into our formula, we obtain

$$P(C_3|M_2) = \frac{\frac{1}{3}(1)}{\frac{1}{3}P(M_2|C_1) + \frac{1}{3}(0) + \frac{1}{3}(1)} = \frac{1}{P(M_2|C_1) + 1}.$$

(Notice, incidentally, that if we had set out to evaluate $P(C_1|M_2)$ instead of $P(C_3|M_2)$, then the unknown term $P(M_2|C_1)$ would have appeared twice instead of just once. That is why I opted to do things the way I have. Yeah, that's me. Always thinking!)

Our conclusion is that if we assume the doors are equally likely to conceal the car at the start of the game and that Monty never opens your initial choice or the door with the car, then the updated probability of door three depends on the value of $P(M_2|C_1)$. That is, with what probability does Monty open door two in those cases where the car is behind door one (which we are assuming is your initial choice)?

In version one of the game, we stipulated that Monty chooses his door randomly when he confronts more than one option. This leads to

$$P(M_2|C_1) = \frac{1}{2}.$$

Plugging that value into our formula gives us

$$P(C_3|M_2) = \frac{2}{3},$$

just as we computed previously. On the other hand, in later sections we will consider certain exotic procedures Monty might follow in selecting his door in those cases where he has more than one option. Those scenarios lead to different values of $P(M_2|C_1)$, and consequently to different values of $P(C_3|M_2)$.

One thing is certain, however. Since $P(M_2|C_1)$ is certainly between 0 and 1 inclusive, we see that $P(C_3|M_2)$ must lie between $\frac{1}{2}$ and 1. This means you can never hurt your chances by switching. It also means that only if $P(M_2|C_1) = 1$ (which might happen, for example, if Monty follows a policy of always opening the lowest-numbered goat-concealing door different from the one you initially chose) are the remaining two doors equally likely.

What changes when we consider version two? We still have that the doors are equally likely to begin with, and therefore that

$$P(C_1) = P(C_2) = P(C_3) = \frac{1}{3}.$$

However, it is no longer certain that Monty will open a goat-concealing door. We know only that Monty will not open the door we initially chose, which we are stipulating to be door one, and that he otherwise chooses randomly. We conclude that Monty will now open door two with probability $\frac{1}{2}$. Keeping in mind that M_2 is the event "Monty opens door two and reveals a goat," we have

$$P(M_2|C_1) = \frac{1}{2}, \quad P(M_2|C_2) = 0, \quad P(M_2|C_3) = \frac{1}{2}.$$

Plugging everything into Bayes' theorem now gives

$$P(C_3|M_2) = \frac{\frac{1}{3}\left(\frac{1}{2}\right)}{\frac{1}{3}\left(\frac{1}{2}\right) + \frac{1}{3}\left(\frac{1}{2}\right)} = \frac{1}{2},$$

just as we expected.

In light of this result, be wary of people who say that Monty cannot change the $\frac{1}{3}$ probability of your door simply by revealing a goat. So long as there is a nonzero chance of Monty revealing the car, he can indeed change that probability.

3.9. What If Monty Is Not Always Reliable?

You might be wondering why we are working so hard to solve version two when it would seem there is a simpler option available to us. After you select door one, there are three possibilities, all equally likely. You have chosen either the car, goat one, or goat two. Let us suppose that you have chosen the car. Monty will now choose a door randomly and reveal a goat. You will lose by switching. This scenario plays out with probability $\frac{1}{3}$.

Now suppose you have chosen goat one. This happens with probability $\frac{1}{3}$. Monty will now choose his door randomly, half the time revealing a goat and half the time revealing the car. If he reveals the car, then the game ends. If he reveals the goat, then you will win by switching. Each of these scenarios occurs with probability $\frac{1}{6}$.

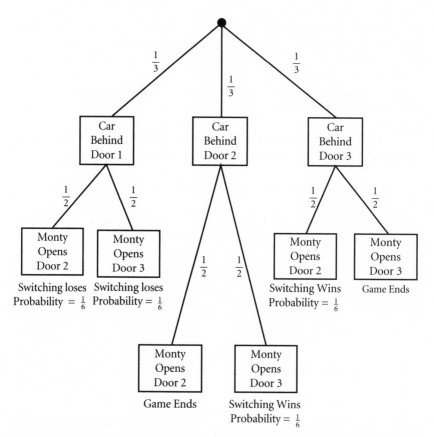

Figure 3.1: Probability tree for version two of the problem when the player initially chooses door one

The reasoning is nearly identical if you have chosen goat two. We again find that in this scenario the game will end prematurely in half of the cases, while in the other half you will win by switching.

Putting everything together shows that there is a probability of $\frac{1}{3}$ that you will lose by switching, a probability of $\frac{1}{3}$ that you will win by switching, and a probability of $\frac{1}{3}$ that the game will end prematurely with Monty revealing the car. If we restrict our attention solely to the cases where Monty does not reveal the car, then we find that you will win half the time by switching and lose the other half of the time. This is precisely what we found previously.

Alternatively, we could have worked out the tree diagram for the problem, as recorded in Figure 3.1.

Thus we have two elementary solutions based entirely on machinery developed in the previous chapter.

Well, yes, we could have solved things that way. We would have saved ourselves quite a bit of effort by doing so. That notwithstanding, there is a payoff for all of our hard work. The machinery of conditional probability can be immediately brought to bear on a variety of Monty Hall variants, as we shall now see.

Version Three: As before, you are shown three equally likely doors. You choose door one. Monty now points to door two but does not open it. Instead he merely tells you that it conceals a goat. You know that in those cases where the car really is behind door one, Monty chooses randomly between door two and door three. You also know that when the car is behind door two or door three, it is Monty's intention to identify the empty door, but that his assertions regarding the location of the car are only correct with probability p, with $p \geq \frac{2}{3}$. What should you do now?

We should pause to note the reason for assuming $p \geq \frac{2}{3}$. We know that two-thirds of the time the car is indeed not behind door two. Consequently, even if Monty is maximally confused and he is really just guessing about the location of the car, we know his statement cannot be wrong with probability greater than $\frac{1}{3}$.

In our previous versions, the primary effect of Monty's actions was the reduction of the probability of one of the doors to 0, thereby leaving you with only two viable options. But now all three doors remain in play. Does that affect our reasoning?

Bayes' theorem makes short work of this question. Let C_1, C_2, and C_3 denote the events that the car is behind door one, two, or three respectively. Let D_2 be the event that Monty asserts door two is empty. We seek $P(C_1|D_2)$.

Evaluating this probability via Bayes' theorem requires that we determine $P(C_1)$, $P(D_2|C_1)$ and $P(D_2)$. The first two terms are easily evaluated, and we obtain

$$P(C_1) = \frac{1}{3}, \quad P(D_2|C_1) = \frac{1}{2}.$$

The second equation follows from our assumption that Monty chooses randomly from among his options when door one conceals the car.

We will evaluate $P(D_2)$ using the law of total probability. We write

$$P(D_2) = P(C_1)P(D_2|C_1) + P(C_2)P(D_2|C_2) + P(C_3)P(D_2|C_3).$$

We certainly have

$$P(C_1) = P(C_2) = P(C_3) = \frac{1}{3}.$$

Now, if the car is actually behind door two, then Monty will wrongly say that it is not with probability $1 - p$. Similarly, when the car is actually behind door three, Monty will correctly say the car is not behind door two with probability p. Therefore:

$$P(D_2|C_2) = 1 - p, \quad P(D_2|C_3) = p.$$

Plugging everything in now leads to

$$P(D_2) = \frac{1}{3}\left(\frac{1}{2} + p + (1 - p)\right) = \frac{1}{2}.$$

This is already an intriguing result. The probability that Monty will point to door two is independent of the value of p.

Now Bayes' theorem reveals that

$$P(C_1|D_2) = \frac{\left(\frac{1}{3}\right)\left(\frac{1}{2}\right)}{\frac{1}{2}} = \frac{1}{3}.$$

It follows that after Monty makes his assertion, we have the following probabilities:

$$P(C_1) = \frac{1}{3}, \quad P(C_2) = 1 - p, \quad P(C_3) = \frac{3p - 1}{3},$$

the last equation following from the requirement that the sum of the three probabilities must be one. From this we conclude that if $p > \frac{2}{3}$, then there is an advantage to be gained from switching doors. If $p = \frac{2}{3}$, then all three doors have the same probability.

It is useful, after carrying out a calculation of this sort, to check your answer for plausibility. We might begin by noticing that the case $p = 1$ corresponds to version one of the Monty Hall problem. This is the case where it is certain that Monty will reveal an empty door. Happily, the formulas above show that door two now has probability 0, while doors one and three have probabilities $\frac{1}{3}$ and $\frac{2}{3}$ respectively. Just as we would expect.

The case $p = \frac{2}{3}$ corresponds to the case where Monty chooses his door randomly. This is just like version two, save for the fact that this time Monty does not actually open his door. As expected, our formulas tell us that the three doors should now be regarded as equally likely. If Monty is choosing randomly, then his assertion gives you no information relevant to determining the location of the car.

3.10. What If the Car Is Not Placed Randomly?

Version Four: Suppose the car is not placed randomly behind the three doors. Instead, the car is behind door one with probability p_1, behind door two with probability p_2, and behind door three with probability p_3. You are to choose one of the three doors, after which Monty will open a door he knows to conceal a goat. Monty always chooses randomly from among his options in those cases where your initial choice is correct. What strategy should you follow?

We can assume, of course, that $p_1 + p_2 + p_3 = 1$. We will further assume, without loss of generality, that $p_1 \geq p_2 \geq p_3$.

Now we have a new wrinkle. Since the car is not placed randomly behind the three doors, we can no longer simplify our work by assuming you initially choose door one, followed by Monty doing something with door two.

What strategy should we follow? Perhaps we should choose door one and stick with it, since it is the most likely to be correct. Or maybe we should choose door three initially, thereby forcing Monty to open a higher-probability door. We will have to investigate several strategies before determining the best course.

We begin by noting there are only six possible strategies. We can pick any of the three doors and stick, or pick any of the three doors and switch.

Our previous experience suggests that since it is assured that Monty will open an empty door, his actions will provide us with no reason for altering the probability of our initial choice. Since our previous experience also suggests the peril of trusting our intuition in anything related to the Monty Hall problem, we may as well consult Bayes' theorem to be certain.

Let us continue to denote by C_i and M_j, for $1 \leq i, j \leq 3$ and $i \neq j$, the events respectively in which door i conceals the car and in which Monty opens door j after we make our initial choice. Assume that our initial choice is door i. Then we have

$$P(C_i|M_j) = \frac{P(C_i)P(M_j|C_i)}{P(M_j)}.$$

This time we have $P(C_i) = p_i$ and $P(M_j|C_i) = \frac{1}{2}$. As always, we will use the law of total probability to compute $P(M_j)$. Note that if we play the game multiple times, then in those cases where our initial choice is incorrect Monty will open each of the other doors in roughly half of the cases. We now have that:

$$P(M_j) = P(C_i)P(M_j|C_i) + P(\bar{C}_i)P(M_j|\bar{C}_i)$$
$$= (p_i)\left(\frac{1}{2}\right) + (1 - p_i)\left(\frac{1}{2}\right) = \frac{1}{2}.$$

Plugging into Bayes' theorem now reveals that

$$P(C_i|M_j) = \frac{p_i\left(\frac{1}{2}\right)}{\frac{1}{2}} = p_i.$$

It is comforting to know our intuitions are not *always* wrong. Our calculation shows that the probability of our initial choice does not change when Monty reveals an empty door. It follows that the optimal sticking strategy is to choose door one and stick with it. (In cases where one or both of p_2 and p_3 is equal to p_1, then of course there may be other sticking strategies that perform equally well.) Thus, by sticking with our initial choice our maximum probability of winning is p_1.

Our calculation has a further useful consequence. Let us suppose we have chosen door i. We know that before Monty opens a door, the two doors different from your initial choice have a joint probability of $1 - p_i$ of containing the car. We have just shown that after Monty opens a door, your initial choice still has probability p_i. The door Monty opens, of course, now has probability 0. This implies that the one other remaining door has probability $1 - p_i$. To maximize our chance of winning with a switching strategy, we should initially choose the door that makes p_i as small as possible. This is accomplished by initially selecting door three and then switching. In this case we win with probability $1 - p_3$.

If we are determined to stick and initially choose door one, then we will win with probability p_1. If we switch and initially choose door three, then we will win with probability $1 - p_3$. Thus, sticking is optimal only if $p_1 \geq 1 - p_3$. Remember, though, that $p_1 + p_2 + p_3 = 1$ and $p_1 \geq p_2 \geq p_3$. It follows that the only way to have $p_1 \geq 1 - p_3$ is to set $p_1 = 1$ and $p_2 = p_3 = 0$. In this case, switching from door three still wins with probability 1, and it makes no difference which of the two strategies we follow. In any other case we gain an advantage by switching.

3.11. Deal or No Deal

As I write this in April 2008, one of the highest-rated programs on American television is the prime-time game show *Deal or No Deal*. Contestants are confronted with twenty-six identical briefcases. Each case contains an amount of money between 1¢ and $1 million. The contestant chooses one of the cases, and this case is placed to one side, unopened.

Play now proceeds as follows: The contestant selects six cases from the remaining twenty-five. These cases are opened and their contents revealed. The play is now paused, and the contestant receives a monetary offer for his case. If the player accepts the offer, then the game is over. If the offer is rejected, then play resumes. The idea is that if the amounts of money revealed in the six opened cases are all very high, then the probability that the contestant's case holds a large sum goes down. But if the six cases all contained small amounts of money, then this probability goes up. Thus, if the contestant has primarily revealed small amounts of money, then he is likely to receive a large offer for his case. If primarily large sums have been revealed, his offer will go down accordingly.

Assuming the initial offer is turned down, the contestant now selects and opens five more cases. Play is again paused so that the contestant may receive an offer. Assuming this is rejected, play resumes with the contestant now selecting four more cases. (Each time an offer is made to the contestant the host asks portentously, "Deal...or no deal?" Hence the name of the show.)

Play continues in this fashion, with offers coming more quickly as the number of cases decreases. Eventually a point is reached where the contestant receives an offer after each new case opening. A typical scenario at this stage might involve five cases remaining, with one containing a large sum of money like $750,000, while the others contain amounts of money under $50,000. The contestant might now receive an offer of $100,000. The contestant might reason that there is an 80% chance of opening something other than the $750,000, and the next offer will go up if he opens anything else. Some balance between risk and reward must now be struck.

If the contestant persistently refuses all offers, a situation arises in which only two cases remain; the originally chosen case and one other. At this point the contestant is given the opportunity of either sticking with her original case or switching to the one remaining unopened case. After making this decision, the contestant's case is opened and she receives the amount of money contained therein. For concreteness, let us assume that the only cases still in play are the ones containing $1 and $1 million. What should the contestant do?

The similarities between *Deal or No Deal* and the Monty Hall problem are obvious. The question is whether this situation more resembles version one or version two.

We might be tempted to argue as follows: When I chose my original case, the probability was only $\frac{1}{26}$ that it contained the $1 million. It follows that the one other remaining case contains the $1 with probability $\frac{25}{26}$. I dramatically increase my chances of winning by switching cases.

This argument is not correct. That *something* is wrong can be seen from the following counterargument: Suppose you do not want the $1 million. For whatever reason, you prefer the $1. As before, you reason that there is a $\frac{1}{26}$ chance that your case contains the $1 and a $\frac{25}{26}$ chance that the other case contains the $1. Thus, you gain a large advantage by switching. It would seem that regardless of what you want, you increase your chances of getting it by switching. Since this cannot be correct, there is plainly a flaw in our argument. A more rigorous argument should help bring it to light.

Once again the problem comes down to updating a prior probability in the face of new information. If A is the event that your initial choice contains the $1 million, then clearly $P(A) = \frac{1}{26}$. Let B denote the event that twenty-four randomly chosen cases from among the remaining twenty-five are opened, and none of them contains the $1 million. We seek $P(A|B)$.

Bayes' theorem handles this situation easily. The only tricky part is determining $P(B)$, since this probability depends in part on the location of the million dollars. Proceeding as before, we write

$$P(B) = P(A)P(B|A) + P(\bar{A})P(B|\bar{A}).$$

We have

$$P(A) = \frac{1}{26}, \quad P(B|A) = 1, \quad P(\bar{A}) = \frac{25}{26}, \quad P(B|\bar{A}) = \frac{1}{25}.$$

Consequently, $P(B) = \frac{2}{26}$. If we now plug everything into Bayes' theorem, we obtain

$$P(A|B) = \frac{P(A)P(B|A)}{P(B)} = \frac{\left(\frac{1}{26}\right)(1)}{\frac{2}{26}} = \frac{1}{2}.$$

Since the probability is now $\frac{1}{2}$ that your case contains the $1 million, there is no advantage to be obtained by switching. The *Deal or No Deal* situation more closely resembles version two of the Monty Hall problem than it does version one. The crucial point is the possibility of revealing the million dollars every time the contestant chooses a case to open. This is analogous to Monty randomly choosing a door to open in version two of the problem.

3.12. The Proportionality Principle

Return now to the classical version of the Monty Hall problem. Let us assume you initially choose door one and Monty then opens door two. We found in Chapter 1 that door three is now twice as likely as door one to conceal the car. We also noted that if the car was behind door one, then it was fifty-fifty whether Monty would open door two or door three; but if the car was behind door three, then Monty was certain to open door two. So the assumption that the car is behind door three makes it twice as likely that Monty will open door two, when compared with the assumption that the car is behind door one. Curious.

We saw a similar principle at work in version two. After choosing door one and seeing Monty open door two, we found that doors one and three were now equally likely to conceal the car. Furthermore, since Monty was now choosing his door randomly, the probability that Monty would open door two was $\frac{1}{2}$ regardless of whether the car was behind door one or door three. Likewise in the *Deal or No Deal* situation. The probability of randomly choosing case X and finding it does not contain the $1 million is the same regardless of the case different from X you assume to contain the $1 million.

There appears to be a general rule here. Writing in [78] and [79], statistician Jeffrey Rosenthal has dubbed it "the Proportionality Principle." He describes it as follows:

> If various alternatives are equally likely, and then some event is observed, the updated probabilities for the alternatives are proportional to the probabilities that the observed event *would* have occurred under those alternatives.

Applying this to version two leads to the following argument: Let us suppose you choose door one and Monty, not knowing where the car is, opens door two. The car is not there. Initially the car was equally likely to be behind any of the three doors. The probability that door two is empty is 1 if the car is actually behind doors one or three, and 0 if the car is behind door two. The updated probabilities for the two remaining doors must therefore preserve the ratio 1:1. It follows that the three doors have probabilities of $\frac{1}{2}$, 0, $\frac{1}{2}$, precisely as we calculated before.

The same logic applies to the classical version. Say we chose door one and Monty opens door two, showing it is empty. The probability that Monty would open door two is $\frac{1}{2}$ if the car is behind door one, 0 if the car is behind door two, and 1 if the car is behind door three. It follows that our updated probabilities for doors one and three must remain in the ratio 1 : 2, which leads to values of $\frac{1}{3}$ and $\frac{2}{3}$ respectively.

Rosenthal's proportionality principle is a straightforward consequence of Bayes' theorem. Let the equally likely alternatives be A_1, A_2, ..., A_n. We will assume that these events occur with nonzero probability. Let B be the event that is causing us to reconsider our probability assignments.

Since we are assuming B has occurred, we can assume $P(B) > 0$. According to Bayes' theorem, we have

$$P(A_i|B) = \frac{P(A_i)P(B|A_i)}{P(B)}.$$

Since we are also assuming the A_i's are equiprobable, we have $P(A_1) = P(A_2) = \ldots = P(A_n)$. It follows that $P(A_i)/P(B)$ is a constant. Calling this constant K, we have

$$P(A_i|B) = K P(B|A_i),$$

which implies that the ratio of the prior probabilities of A_i and A_j is equal to the ratio of their posterior probabilities given the occurrence of B. This is precisely the content of the proportionality principle.

Rosenthal goes on to consider a further variation of the Monty Hall problem. Let us have a look:

> *Version Five: As always, you choose one of three equally likely doors. Monty then opens a door he knows to be empty. This time, however, we assume that Monty opens the lowest-numbered door available to him with probability p (and therefore picks the higher numbered door with probability $1 - p$). Is there any advantage to switching doors in this case?*

We will assume throughout that you initially choose door one. Let us suppose first that Monty opens door two. We now have three possibilities.

- If the car is behind door one, then Monty will open door two with probability p.

- If the car is behind door two, then Monty will not open that door.

- If the car is behind door three, then Monty opens door two with probability 1.

Using the proportionality principle, we would conclude that the updated probabilities for door one and door three must maintain the ratio $p : 1$. This implies that the probability of door one is now $\frac{p}{1+p}$, and the probability of door three is now $\frac{1}{1+p}$. If $p = 1$, meaning that Monty is certain to open the lowest-numbered door available to him, then we find that doors one and three are now equiprobable. In any other scenario there is an advantage to switching doors. In particular, the case $p = \frac{1}{2}$ corresponds to version one of the Monty Hall problem, and our formulas quite properly tell us that door one now has probability $\frac{1}{3}$, while door three has probability $\frac{2}{3}$.

The reasoning is similar if Monty opens door three instead. For now we would reason that if the car is behind door one, then Monty opens door three with probability $1 - p$, while if the car is behind door two, he opens door three with probability 1. (If the car is behind door three, then there is probability 0 that Monty will open that door, of course.) So the updated probabilities for doors one and two must maintain the ratio $1 - p : 1$. Consequently, door one now has probability $\frac{1-p}{2-p}$, while door two now has probability $\frac{1}{2-p}$. Note once again that these formulas give the correct answer in the case where $p = \frac{1}{2}$, which corresponds to version one of the problem.

3.13. Interpretations of Probability

We have been discussing probability for more than a few pages now, but still we have not answered a fundamental question. What are probability statements, *really*?

When I say a tossed coin will land heads with probability $\frac{1}{2}$, should that be interpreted as a statement about coins? Or is it instead a statement describing my level of ignorance regarding the outcomes of coin tosses? Does the probability of an event depend solely on the event itself, implying that it may be discovered via judicious experimentation? Or is it something that changes depending on the information we have regarding the event?

Both views have a certain plausibility. Fair coins flipped in a fair manner really do come up heads roughly half the time when tossed a large number of times. That is as much a fact about coins as their mass or their metallurgical makeup. On the other hand, an understanding of elementary physics tells us that the coin is not really subject to the whims of chance at all. Its entire trajectory is determined by the laws of Newtonian mechanics from the moment it leaves my hand, and if I were in possession of all the relevant facts, I could, at least in principle, predict with certainty how the coin will land. Seen in that way, how can a probability assignment represent anything more than a description of my personal level of ignorance?

In pondering such questions, we leave behind the rolling meadows and soothing greenery of pure mathematics and enter instead the jagged cliffs and icy darkness of philosophy. Mathematicians, you see, are perfectly content to treat probability theory as an abstract construction no different from anything else they study. Just as Euclid laid down at the start certain unproved axioms and undefined notions that would form the basis for his work in geometry, so too do high-level texts on probability theory begin by laying down the basic rules by which all probability measures must play. What is a probability measure? It is a function mapping certain kinds of sets to real numbers in such a way that various pleasing assertions are seen to hold. That is all. From here you are constrained only by the principles of logic, and may commence proving theorems with all haste.

And if conceptual difficulties arise when trying to assign actual real-world meaning to those sets and functions? Well, that just serves you right for trying to apply them to anything.

That was how I viewed things before starting work on this book. It is, to a large extent, how I still view things. However, I was moved to reconsider this question when I reread what I had written to this point and discovered that I had, without even realizing it, made free use of three different interpretations of probability.

Our first notion was based on the idea of an experiment with several, equally likely, possible outcomes. Given this setup, the probability of an event was defined as the ratio of favorable outcomes to possible outcomes. This is known as the classical interpretation of probability, in honor of its historical significance. This approach seems well suited to questions arising from gambling, and since the earliest thinkers in this subject were motivated by such questions, it is unsurprising that they arrived at this definition.

Having employed the classical interpretation in our treatment of the basic Monty Hall scenario, we then sought empirical confirmation of our results. These were obtained via a Monte Carlo simulation. The usefulness of such simulations suggests a frequentist interpretation of probability. In this view a probability is something you measure from actual data. The law of large numbers plays a central role here, since it provides the justification for thinking that data collected from a long run of trials are a good proxy for the actual probability whose value is sought. Since a reliance on finite collections of data implies that probabilities keep changing as more data are collected, it is often argued that probabilities represent the limiting frequencies of an infinite number of trials.

The present chapter has seen something else entirely. In several places we used probabilistic language to describe our degree of confidence in a given proposition. It is difficult to discuss applications of Bayes' theorem without taking this view. Thus, in describing the problem of the baggage carousel at the airport, the issue was the probability that ought to be assigned to the proposition "Our bag has been misplaced by the airline" in the face of the evidence of the number of bags that had already appeared. In solving the problem we were not attempting to learn something about luggage or airport procedure, but instead were trying to decide what is rational to believe regarding the fate of our luggage. This is referred to as the Bayesian interpretation of probability.

A major sticking point between these interpretations is whether probability assignments are objective or subjective. The frequentist interpretation is an objective view of things. Probabilities are things you learn about by collecting data from multiple runs of a repeatable experiment. Bayesianism takes the opposite view. Probabilities represent the degree of confidence a person has in a proposition, and it is something that can, and indeed should, be changed as new information comes in. The classical view lies somewhere in between these extremes. It is subjective in the sense that classical probabilities are assigned not with regard to actual data but rather with regard to our understanding of some particular experiment. On the other hand, probability assignments are seen as descriptive of what will happen if a long run of trials is carried out. Indeed, the law of large numbers establishes the proper relationship between classical probabilities and measured frequencies in long runs of trials.

Perhaps a quick example will illuminate the differences. To the classicist, the assignment of $\frac{1}{2}$ to the probability of tossing a coin and having it land heads means that there are two equally likely outcomes, only one of which is heads. To the frequentist it means that in a very large number of tosses we expect to get heads roughly half the time. To the Bayesian we are saying merely that the information we have does not justify the conclusion that either heads or tails is more likely than the other.

Suppose we now perform a Monte Carlo simulation in which a coin is tossed 10,000 times and 5,043 heads are obtained. The classical theorist would see the measured frequency of heads as sufficiently close to $\frac{1}{2}$ to justify the conclusion that his probability assignment was correct. The frequentist would compute a probability of 5,043/10,000 for the probability of heads in this collection of data, and argue that as the number of trials approached infinity we would see, with very high probability, the measured frequency of heads get closer and closer to $\frac{1}{2}$. The Bayesian would see a new piece of evidence regarding the fairness of the coin, and he would use this evidence to update (and most likely confirm) his belief that the coin was fair.

The philosophical literature abounds with point and counterpoint regarding these interpretations, and several others besides. The classical interpretation certainly seems well suited to fairly simple cases, where exhaustive lists of outcomes are readily enumerated and equiprobability can be recognized when it presents itself. On the other hand, it suffers from a circularity at its core. It defines probability as the ratio of favorable outcomes to possible outcomes, but only when the outcomes have already been deemed to be equiprobable, which is distressingly close to the term we were trying to define in the first place. Furthermore, once elementary examples drawn from casino games are left behind and more complex problems attempted, its practical utility turns out to be severely limited.

Frequentist views certainly capture much of our intuitive understanding of probability. The law of large numbers, as Jakob Bernoulli has informed us, is something everyone understands without needing to be told. That certain chance scenarios lead to stable long-run frequencies can hardly be denied, and probability language seems well suited for describing these situations. Furthermore, there is an extensive literature among psychologists and cognitive scientists suggesting that people are natural frequentists. For example, studies have shown that when the Monty Hall problem is presented in frequentist terms (as opposed to an exercise in decision theory, as we have presented it), the confusion about the problem diminishes greatly. We will delve into this literature in Chapter 6. In light of this, it is unsurprising that frequentist views tend to dominate among scientists and statisticians.

Philosophers take a different view, generally regarding frequentist interpretations as inadequate. Serious conceptual difficulties arise from trying to equate probability with relative frequency. It starts with simple worries. If we define the probability of X to be the ratio of X occurrences to all occurrences in a suitable long run of trials, then we are committed to the idea that probabilities must be rational numbers. Physicists, alas, report that irrational probabilities are indispensable in the study of quantum mechanics. It does not help to define the probability of X to be the limiting frequency of X in an infinity of trials (which would allow irrational probabilities to be realized as the limit of a sequence of rational numbers). For then I would be forced to point out that there is no clear reason for thinking that such limits even exist. No one can actually carry out infinitely many trials, after all, and data from any finite number of trials are ultimately consistent with any conceivable limiting behavior.

There is also the reference class problem. Suppose I wish to determine the probability that a particular forty-year-old male will develop lung cancer. My male subject may be classified among males generally, among those of the same nationality, among those in the same line of work, among those with similar eating habits, and so on endlessly. I could compute a relative frequency of lung cancer sufferers among males in each group, but which is the correct probability? It would seem that frequency by itself cannot be the whole story, but instead must be considered in the context of a properly chosen reference class. In certain situations we might say that one particular reference class is plainly more appropriate than the others, thereby avoiding the problem. Suffice it to say that in many situations this is not the case.

And then there is the most basic problem of all. How do I handle situations where it is not possible to carry out multiple trials of an experiment? For example, what is the probability that my friend will marry in the next year? What is the probability that a particular medical research project will yield results sufficient to justify the cost? These seem like perfectly reasonable instances of probability language, but they do not fit well with a frequentist interpretation.

What of the Bayesians? Certainly probability as degree of belief has a lot going for it, and this general approach does seem to be more widely applicable than either classicism or frequentism. This wide applicability is purchased at great cost, however. For one thing, most people find the Bayesian view to be needlessly complex and counterintuitive. If it amuses you to treat the observation that heads came up 49.6% of the time in a large number of coin tosses as a piece of evidence that gets plugged into Bayes' theorem for the purpose of updating a prior probability assignment, then more power to you. But I am not the crazy one for seeing a fact about coins in the same observation. A further problem is that Bayesianism does not get off the ground until a prior probability distribution is defined, but it is not clear from where this distribution ought to come. In certain extreme views an agent is deemed rational so long as his probability assignments do not contradict the rules of the probability calculus (for example, he must not assign a probability of 3 to anything). Thus, you are rational in assigning a probability of 1 to the proposition "The moon is made of green cheese" so long as you are then careful to assign a probability of 0 to the proposition "The moon is not made of green cheese." Others, finding this too extravagant, require that our probability assignments be subject to the whims of an "ideal reasoner." This opens a different can of worms.

Believe me when I tell you that I have barely scratched the surface. Thick, impenetrable books get written on this subject, from a variety of perspectives, and I never would have given any of it a second look were it not for the Monty Hall problem. It would seem the problem forces you to consider not just the mathematical aspects of probability but the philosophical aspects as well.

In the end my reaction to what I have read is typical of my reaction to philosophical literature generally. It is subtle, nuanced, absolutely ingenious, and ultimately just not terribly important. We are talking, after all, about interpretations of probability. The mathematical formalism is agreed to by all parties. At issue is simply what it all means in any given practical context. Surely the primary criteria here are usefulness and correctness. For that reason I prefer the ecumenical view taken by most mathematicians. If you find it helpful to think of probability in a particular way, and if your preferred interpretation leads you to true statements about whatever it is that you are studying, then you go right ahead and adhere to that interpretation. Why is it even reasonable to think that a single interpretation can capture every facet of probability?

The most natural approach to the basic Monty Hall problem involves the classical view of probability. In seeking empirical verification for our solution, it was best to think in frequentist terms. And in analyzing more complex host behaviors, a Bayesian view was indispensable. Three different interpretations useful in three different contexts. Exactly as it should be.

4

Progressive Monty

4.1. Proceed at Your Own Risk

Version Six: This time we assume there are n identical doors, where n is an integer satisfying $n \geq 3$. One door conceals a car, the other $n - 1$ doors conceal goats. You choose one of the doors at random but do not open it. Monty then opens a door he knows to conceal a goat, always choosing randomly among the available doors. At this point he gives you the option either of sticking with your original door or switching to one of the remaining doors. You make your decision. Monty now eliminates another goat-concealing door (at random) and once more gives you the choice either of sticking or switching. This process continues until only two doors remain in play. What strategy should you follow to maximize your chances of winning?

If you are not yet convinced that the Monty Hall problem is little more than a relentless assault on your probabilistic intuition, then this version ought to persuade you. We will refer to it as the progressive Monty Hall problem.

I first encountered it in the paper by Bapeswara Rao and Rao [77]. There they specifically considered the case of four doors and concluded, by enumerating the sample space, that the optimal strategy is to stick with your initial choice until only two doors remain, and then switch. We will consider their argument in detail in the next section.

The provocative part of the essay was the assertion that this "switch at the last minute" (SLM) strategy was optimal for any number of doors. This is plausible. After all, switching doors for the first time when only two doors remain is precisely what we do in the classical version of the problem. The SLM strategy can be seen as a straight generalization of what we already know. Still, it would be nice to have a proof. It is not at all clear why switching somewhere in the middle of the game necessarily harms your chances of winning. Since listing all the possibilities is not practical when the number of doors is large, we will have to look elsewhere for a proof.

We will, eventually, provide two detailed arguments for preferring the SLM strategy over all rivals, but I should warn you at the outset that we have our work cut out for us. The technical details will be considerably more formidable in this chapter than they have been previously.

4.2. The Four-Door Case

As a warm-up exercise, let us consider the four door case. We will follow closely the approach in [77].

To keep our analysis to a reasonable length, we will consider only the following three strategies:

1. Select a door at random and stick with it throughout.

2. Switch doors at every opportunity, choosing your door randomly at each step.

3. Stick with your first choice until only two doors remain, and then switch.

Strategy one is readily analyzed. Our experience thus far tells us that since Monty is guaranteed to open an empty door and chooses randomly from among his options at every step, the probability of our initial choice cannot change if we stick with it. Consequently, by following strategy one you win only when your initial choice is correct, and this happens with probability $\frac{1}{n}$. Pretty grim.

Strategy three corresponds to the SLM approach. The authors of [77] give the following argument for thinking that you win with probability $\frac{3}{4}$ by following it: Denote by A the event in which your initial door choice conceals the prize, and by $\bar{A}$ the event in which your initial choice does not conceal the prize. Then we have $P(A) = \frac{1}{4}$ and $P(\bar{A}) = \frac{3}{4}$. Since Monty only opens empty doors, we can win with this strategy only if $\bar{A}$ occurs. Consequently, SLM wins with probability $\frac{3}{4}$.

This argument, while correct, leaves something to be desired. It fails to make clear the significance of Monty's inability to open the car-concealing door, and it gives little indication of how to handle variations of the problem in which there is a nonzero probability of Monty revealing the car. And if

Table 4.1: Winning scenarios when switching every time in the four-door progressive game

Car	Monty	You	Monty	Probability
1	2	3	4	1/24
1	2	4	3	1/24
1	3	2	4	1/24
1	3	4	2	1/24
1	4	2	3	1/24
1	4	3	2	1/24
2	3	4	1	1/16
2	4	3	1	1/16
3	2	4	1	1/16
3	4	2	1	1/16
4	2	3	1	1/16
4	3	2	1	1/16

ever there was a situation in which you should not be satisfied with simple, intuitively satisfying arguments, the Monty Hall problem is it. There is also considerable insight to be gained from a detailed analysis of the conditional probabilities that arise from this strategy. For those reasons we will provide a more detailed argument in the next section.

We will now consider the possibility of switching randomly at each step, assuming (without loss of generality) that you start with door one. There are then twelve scenarios in which you win with this strategy. We list them in Table 4.1, together with the probability of each scenario. The first column records the location of the car. The following three columns represent the first door Monty opens, the door you switch to, and the second door Monty opens respectively. Since we are only listing winning scenarios, your final choice must match the location of the car in each case.

If you are uncertain as to how the probabilities were calculated, take a look at the first row of the table. In that scenario the car is assumed to be behind door one, which happens with a probability of $\frac{1}{4}$. Since that is also the door we have chosen, Monty can then open any of the remaining three doors. Therefore, the probability that he opens door two is $\frac{1}{3}$. You will now switch at random to either of the remaining two doors, choosing door three with probability $\frac{1}{2}$. Monty is now forced to open door four. It follows that this scenario occurs with probability

$$\frac{1}{4} \times \frac{1}{3} \times \frac{1}{2} \times 1 = \frac{1}{24}.$$

Now look at row seven. This time the prize is behind door two, which happens with probability $\frac{1}{4}$. But this time Monty can open neither door one (because it conceals the prize) nor door two (because it is your current choice). Consequently, he must choose between door three and door four, and will choose the former with probability $\frac{1}{2}$. You now choose door four with probability $\frac{1}{2}$, after which Monty is forced to open door three. Therefore, we obtain a probability of

$$\frac{1}{4} \times \frac{1}{2} \times \frac{1}{2} \times 1 = \frac{1}{16}.$$

If we now add up the probabilities of all the scenarios given in the table, we find that we win with probability $\frac{5}{8}$ by following a strategy of switching every time. This is smaller than the $\frac{3}{4}$ we obtained via SLM.

We could provide a similar analysis of scenarios in which we switch exactly once somewhere in the middle. The analysis is not difficult, but it is quite tedious. We will therefore omit the details and content ourselves with the simple observation that all such strategies are inferior to SLM.

It seems, therefore, that SLM is the uniquely optimal strategy in the four-door progressive Monty Hall problem. Is this still true for n doors?

4.3. Switching at the Last Minute

We now consider the case of an arbitrary number of doors. Our quest to prove that SLM is uniquely optimal begins by establishing that it offers a probability of $\frac{n-1}{n}$ of victory. We could do that with an argument analogous to the one discussed in the previous section for the four-door case. However, for the reasons mentioned there, it is probably best to provide a more detailed analysis. In fact, we will offer two solutions to the problem.

4.3.1. The Direct Attack

We can assume without loss of generality that you initially choose door one. This door has probability $\frac{1}{n}$ at this time. Monty now opens door i, with $2 \leq i \leq n$, to reveal a goat. We must show that this revelation provides no reason for altering the probability of door one. We will continue to denote by C_i the event that the car is behind door i and by M_i the event that Monty opens door i.

According to Bayes' theorem, we have

$$P(C_1|M_i) = \frac{P(C_1)P(M_i|C_1)}{P(M_i)}.$$

Since Monty chooses randomly from among the $n - 1$ goat-concealing doors, and since the doors are assumed to be identical initially, we have

$$P(M_i|C_1) = \frac{1}{n-1} \quad \text{and} \quad P(C_i) = \frac{1}{n}.$$

It remains to determine $P(M_i)$. The probability that Monty opens door i depends on the location of the car. We will recognize three possibilities:

the car is either behind door one, behind door i, or it is somewhere else.

- If the car is behind door one, which happens with probability $\frac{1}{n}$, then Monty can open any of the remaining $n - 1$ doors with equal probability.
- If the car is behind door i, then there is a probability of 0 that Monty will open that door.
- If the prize is anywhere else (which happens with probability $1 - \frac{2}{n}$ or $\frac{n-2}{n}$), then Monty will open any of the $n - 2$ doors different from door one and door i with equal probability.

These observations lead to the following computation:

$$P(M_i) = P(M_i|C_1)P(C_1) + P(M_i|C_i)P(C_i) + P(M_i|\bar{C}_1 \cap \bar{C}_i)P(\bar{C}_1 \cap \bar{C}_i)$$

$$= \frac{1}{n-1}\left(\frac{1}{n}\right) + 0\left(\frac{1}{n}\right) + \frac{1}{n-2}\left(1 - \frac{2}{n}\right) = \frac{1}{n-1}.$$

(Note that in general we have $P(M_i|C_i) = 0$. If the car is behind door i, then Monty will not open door i. We will omit this term from future calculations of this sort.)

Plugging everything into Bayes' theorem now reveals that

$$P(C_1|M_i) = \frac{\left(\frac{1}{n}\right)\left(\frac{1}{n-1}\right)}{\frac{1}{n-1}} = \frac{1}{n},$$

and we conclude that Monty's actions have provided no reason for altering the probability of door one.

We now use this observation as the base case of a proof by induction. Assume that Monty has eliminated x doors, with $0 \leq x \leq n - 3$. (The case $x = 0$ is what we have considered thus far.) Since we are following SLM, we assume that we have stuck with door one throughout. By the inductive hypothesis, we can assume that door one still has probability $\frac{1}{n}$. Monty will now open another empty door. We will refer to this as door j. We want to show that $P(C_1|M_j) = \frac{1}{n}$. By Bayes' theorem, this is equivalent to showing that $P(M_j|C_1) = P(M_j)$.

Note that there are $n - x - 1$ doors other than door one remaining in play. Since none of these doors has been chosen at any stage of the game, we should regard them as equiprobable with probabilities summing to $\frac{n-1}{n}$. Since Monty chooses randomly from among the empty doors, we can write:

$$P(C_j) = \frac{n-1}{n(n-x-1)} \quad \text{and} \quad P(M_j|C_1) = \frac{1}{n-x-1}.$$

Finally, arguing in a manner analogous to our handling of the base case, we have

$$P(M_j) = P(M_j|C_1)P(C_1) + P(M_j|\bar{C}_1 \cap \bar{C}_j)P(\bar{C}_1 \cap \bar{C}_j)$$

$$= \left(\frac{1}{n-x-1}\right)\left(\frac{1}{n}\right) + \left(\frac{1}{n-x-2}\right)\left(1 - \frac{1}{n} - \frac{n-1}{n(n-x-1)}\right)$$

$$= \frac{1}{n-x-1}.$$

Therefore, we have $P(M_j|C_1) = P(M_j)$, as desired.

It follows that $P(C_1) = \frac{1}{n}$ regardless of the number of doors Monty has opened, so long as we do not change from our initial choice. The effect of Monty's door opening is to redistribute the entire $\frac{n-1}{n}$ probability equally over the remaining doors different from our initial choice. Consequently, by switching when only one other door remains, we will win with probability $\frac{n-1}{n}$.

4.3.2. An Elegant Alternative

There is an alternative proof of our result that merits consideration.

It begins with the observation that, according to Bayes' theorem, the updated probability of C_1 given that the event M_i has occurred (that is, given that Monty has opened some door other than your initial choice) is obtained by multiplying $P(C_1)$ by the fraction $\frac{P(M_i|C_1)}{P(M_i)}$. We will evaluate the terms of this fraction.

Let us assume that we have reached a stage in the game where x doors remain. We assume that we are sitting on door one. Since Monty can open any goat-concealing door different from our present choice and since he selects his door at random, we have

$$P(M_i|C_1) = \frac{1}{x-1}.$$

How do we evaluate $P(M_i)$, where $i \neq 1$ (since Monty will not open our current door choice)? According to the law of total probability, we find the sum of all terms of the form $P(M_i|C_j)P(C_j)$, where $i \neq j$ (because if door i conceals the car then there is a probability of zero that Monty will open that door.) For example, if we are playing the four-door game and have initially chosen door one, then the probability that Monty will now open door three is given by

$$P(M_3) = P(M_3|C_1)P(C_1) + P(M_3|C_2)P(C_2) + P(M_3|C_4)P(C_4).$$

We leave out the term $P(M_3|C_3)P(C_3)$, because it is plainly equal to zero.

The terms $P(M_i|C_j)$ can only take on two values. We have already evaluated the case where $j = 1$. For all other cases we have

$$P(M_i|C_j) = \frac{1}{x-2}.$$

Let us now assume that door one has probability p and that all of the other remaining doors are equiprobable. Since there are $x - 1$ doors different from door one, and since these doors collectively have a probability of $1 - p$, we see that

$$P(C_j) = \frac{1-p}{x-1},$$

for all values of $j \neq 1$.

We can now evaluate the sum for $P(M_i)$. It consists of one term of the form

$$P(M_i|C_1)P(C_1) = \left(\frac{1}{x-1}\right)p,$$

and $x - 2$ terms of the form

$$P(M_i|C_j)P(C_j) = \left(\frac{1}{x-2}\right)\left(\frac{1-p}{x-1}\right).$$

It follows that

$$P(M_i) = \frac{p}{x-1} + (x-2)\left(\frac{1}{x-2}\right)\left(\frac{1-p}{x-1}\right)$$

$$= \frac{1}{x-1} = P(M_i|C_1).$$

This implies that the posterior probability of door one is the same as its prior probability. Monty's door-opening gives us no reason to change our assessment of the probability.

It only remains to note that our assumptions certainly hold at the start of the game. Door one has probability $\frac{1}{n}$ at that time, and all of the other doors have the same probability. Since the unchosen, unopened doors are always equiprobable at any stage of the game, we find that the probability of our initial choice cannot change until we switch to a different door.

4.4. A Case Study

Perhaps you are tempted by the following argument: We have seen that SLM wins with probability $\frac{n-1}{n}$. Now, Monty cannot cause the probability of a door to decrease simply by opening some other door. That is, the knowledge that door x is empty cannot make door y seem less likely to be correct. It follows that the probability we assign to a door upon choosing it for the first time is the smallest probability that door will have for the rest of the game. Our initial choice has probability $\frac{1}{n}$. Any door chosen after the first will have probability greater than $\frac{1}{n}$ at the moment we choose it. Assume now that we are following any strategy other than SLM and fast-forward to the moment when only two doors remain. Since we are not following SLM, we have switched at some point from our original door. It follows that the door we are on has probability greater than $\frac{1}{n}$. The other remaining door must then have probability smaller than $\frac{n-1}{n}$. This implies that whatever strategy we are following wins less frequently than does SLM.

Convinced? I sure was. For a while, anyway. After devising that argument I spent a fair amount of time patting myself on the back for my cleverness. I should have realized, however, that my proof was vulnerable to an equally persuasive counterargument: this is a variation of the Monty Hall problem, and therefore nothing is as it seems.

If this argument were correct, there would be no need for this chapter to go on for as long as it does. Sadly, it is not correct. To see why, let us consider a specific example of the progressive Monty Hall problem in the five-door case. We will use a quintuple of rational numbers to represent the probabilities of the five doors at some stage of the game, and we will refer to this quintuple as the probability vector for that stage of the game.

Of course, as the game begins we have probability vector

$$\left(\frac{1}{5}, \frac{1}{5}, \frac{1}{5}, \frac{1}{5}, \frac{1}{5} \right).$$

Suppose we initially choose door one, and then Monty opens door two. According to our results from the previous section we ought to find that $P(C_1|M_2) = \frac{1}{5}$. Indeed, since

$$P(C_1) = P(C_2) = \frac{1}{5},$$

we compute that

$$P(C_1|M_2) = \frac{P(C_1)P(M_2|C_1)}{P(M_2|C_1)P(C_1) + P(M_2|\bar{C}_1 \cap \bar{C}_2)P(\bar{C}_1 \cap \bar{C}_2)}$$

$$= \frac{\frac{1}{5}\left(\frac{1}{4}\right)}{\frac{1}{4}\left(\frac{1}{5}\right) + \frac{1}{3}\left(\frac{3}{5}\right)} = \frac{1}{5}.$$

Since doors three, four, and five are equiprobable and have probabilities that sum to $\frac{4}{5}$, it follows that we now have the probability vector

$$\left(\frac{1}{5}, 0, \frac{4}{15}, \frac{4}{15}, \frac{4}{15} \right).$$

What happens if we now switch to door three and Monty then opens door five? How does that affect the probability of doors one and three?

In working through the following calculations, it will be useful to keep in mind that there are two kinds of doors that Monty cannot open: the door concealing the car, and the door you have currently chosen. Also keep in mind that Monty always chooses randomly among the doors not answering to either of those descriptions.

We have that

$$P(C_1) = \frac{1}{5}, \quad \text{and} \quad P(C_3) = P(C_4) = \frac{4}{15}.$$

We also have that

$$P(M_5|C_1) = P(M_5|C_4) = \frac{1}{2} \quad \text{and} \quad P(M_5|C_3) = \frac{1}{3}.$$

We can compute $P(M_5)$ using the law of total probability as follows:

$$P(M_5) = P(M_5|C_1)P(C_1) + P(M_5|C_3)P(C_3) + P(M_5|C_4)P(C_4)$$

$$= \frac{1}{2}\left(\frac{1}{5}\right) + \frac{1}{3}\left(\frac{4}{15}\right) + \frac{1}{2}\left(\frac{4}{15}\right) = \frac{29}{90}.$$

Consequently, we can use Bayes' theorem to compute that

$$P(C_1|M_5) = \frac{P(C_1)P(M_5|C_1)}{P(M_5)} = \frac{\frac{1}{5}\left(\frac{1}{2}\right)}{\frac{29}{90}} = \frac{9}{29},$$

and

$$P(C_3|M_5) = \frac{P(C_3)P(M_5|C_3)}{P(M_5)} = \frac{\frac{4}{15}\left(\frac{1}{3}\right)}{\frac{29}{90}} = \frac{8}{29}.$$

This produces the probability vector

$$\left(\frac{9}{29}, \ 0, \ \frac{8}{29}, \ \frac{12}{29}, \ 0\right).$$

An interesting result. The probability of all three remaining doors increased as a result of Monty opening door five. In particular, the probability of door one increased. Had we stuck with door one, the opening of door five would not have altered this probability.

But the real surprise occurs in the following scenario: Suppose that after we switched to door three, Monty opened door one instead of door five. How does that affect the probability of door three?

Using Bayes' theorem and the law of total probability, we can write

$$P(C_3|M_1) = \frac{P(C_3)P(M_1|C_3)}{P(M_1|C_3)P(C_3) + P(M_1|C_4)P(C_4) + P(M_1|C_5)P(C_5)}.$$

This time the appropriate values to plug in are

$$P(C_3) = P(C_4) = P(C_5) = \frac{4}{15}$$

$$P(M_1|C_4) = P(M_1|C_5) = \frac{1}{2}$$

$$P(M_1|C_3) = \frac{1}{3}.$$

Plugging everything in leads to

$$P(C_3|M_1) = \frac{\frac{4}{15}\left(\frac{1}{3}\right)}{\frac{1}{3}\left(\frac{4}{15}\right) + \frac{1}{2}\left(\frac{4}{15}\right) + \frac{1}{2}\left(\frac{4}{15}\right)} = \frac{1}{4},$$

which gives us the probability vector

$$\left(0, \ 0, \ \frac{1}{4}, \ \frac{3}{8}, \ \frac{3}{8}\right).$$

Consequently, the probability of door three goes down, from $\frac{4}{15}$ to $\frac{1}{4}$, when we see Monty open door one. Very surprising! The proof I gave at the start of this section was premised on the idea that a door's probability could not go down when Monty opens some other door. It would seem that this premise is not correct.

The fact that a door's probability can decrease when Monty opens a different door makes it highly nontrivial to prove that SLM really is the best strategy. Perhaps there is some clever sequence of switches and subsequent door openings that drives the probability of at least one door below $\frac{1}{n}$. If such a thing were possible, it is conceivable there is a strategy superior to SLM. Even though we have no control over the doors Monty opens, this possibility cannot be so readily dismissed.

This explains the hard work required to establish the optimality of SLM.

4.5. The Calm Before the Storm

The remainder of this chapter will be very heavy mathematically. It is regrettable but unavoidable. As we have had ample cause to note in getting to this point, intuitive arguments do not cut it in dealing with the Monty Hall problem.

That does not mean, however, that we must abandon intuition altogether. Our ruminations about conditional probability and our case study from the previous section have provided some guidance. Perhaps we can now explain in general terms what is going on.

Here are some basic principles to keep in mind:

1. The effect of choosing a door is to make it impossible for Monty to open that door.

2. When Monty appears to have an opportunity to open a given door but chooses not to, the probability of that door goes up.

3. Monty is more likely to open a low-probability door than he is to open a high-probability door.

4. When Monty eliminates all of the doors that have been previously chosen (other than the current choice), the effect is to erase the prior history of the game.

How does our case study look in the light of these principles? We start by choosing door one, which is correct with probability $\frac{1}{5}$. What sort of information do we receive when Monty now shows us that door two is empty? Obviously, we learn that door two is incorrect. But how does door three look to us now? Why did Monty not open *that* door? One possibility is that door three simply lost the lottery when Monty was selecting his door. Another possibility, however, is that door three conceals the car. It is this possibility that causes us to revise upward our probability assignment for door three. This is precisely the possibility that does not apply to door one. We learn nothing from the fact that Monty failed to open this door, because I removed that door from consideration by selecting it.

After Monty opened door two we switched to door three. Monty opened door five. Our calculations showed that all of our probability assignments were now revised upward. Why is that?

That door four now seems more likely is readily understood. As far as we know, Monty had the option of opening that door, so by choosing not to he has increased our assessment of its probability. Door one can also be understood in those terms. By switching to door three, we made that door available to Monty, so we gain information when Monty chooses not to take advantage of our largesse.

There is more we can say. Prior to Monty's opening door five, we thought door one was the least likely of the doors to conceal the prize. That makes it the door most likely to be opened by Monty, as suggested by principle two. This fact is most readily understood by considering an extreme case. Suppose we are in a situation where only three doors remain, having probabilities of $\frac{1}{100}$, $\frac{2}{100}$, and $\frac{97}{100}$. Imagine we are sitting on the middle door. What do we think will happen next? Monty must open either the low-probability door or the high-probability door. It is near certain that the car is behind the high-probability door. Consequently, we would be very surprised if Monty now opened that door. If Monty does so nevertheless, we would now consider it far more likely than previously that the car is behind the low-probability door. (That the lowest-probability door different from your current choice is the one most likely to be opened can be proved rigorously using the law of total probability, but I will not belabor those details here.)

What of door three? Monty was not allowed to open that door in this round, so the mere fact that it remained closed is non-informative. We might think that Monty's actions give us no reason to revise the probability of door three, just as Monty's first door opening did not alter the probability of door one (our initial choice). The difference between the two cases is that the doors were no longer equiprobable when I switched to door three. At the time I made my switch the doors had been cleanly divided into two groups: door one, which was very unlikely to conceal the car; and doors three, four, and five, which were equiprobable and collectively very likely to conceal the car. Prior to Monty's door opening I thought it was very likely that I would find the car among doors three, four, and five. That fact does not change when Monty opens door five. Consequently, the elimination of door five makes it seem more likely that the car is behind door three or door four.

What happened when Monty opened door one rather than door five? That the probabilities of doors four and five now went up is unsurprising. Since Monty might have opened either of them, it is informative to see that he chose not to do so. But why did the probability of door three go down?

To see what is happening, imagine that you are playing the progressive Monty Hall game starting with four doors. You make a selection and then Monty opens a door. Your initial selection now has probability $\frac{1}{4}$, while the other two doors each have probability $\frac{3}{8}$.

Now let us change things slightly. Again suppose that you are starting with four doors. This time, however, we will assume one of the doors has a very low probability of being correct, while the other three are equiprobable. You choose from among the three, and Monty now eliminates the low-probability door. Does this situation differ at all from the one in the previous paragraph? Do you possess relevant information in one of the scenarios that you do not have in the other? The answer is no, and that is why you assign the same probabilities in both situations. This is what I meant in saying that when Monty eliminates all of the previously chosen doors, he effectively erases the prior history of the game.

To test our newfound intuition, let us go one more round with our case study. After we chose door one, Monty opened door two, we switched to door three, and Monty opened door one, the probability vector was

$$\left(0, 0, \frac{1}{4}, \frac{3}{8}, \frac{3}{8}\right).$$

What happens if we now switch to door four? Suppose Monty opens door five. Door three should have its probability go up. As the lowest-probability door, it was the one most likely to be opened. That Monty opened door five instead makes door three look more likely to conceal the car. Door four should likewise have its probability go up, since it was among a proper subset of equiprobable options whose numbers have now been reduced. As for which of the doors now seems more likely, I fear that is not something that can be resolved by intuition alone. Only an actual calculation will tell us how to weight the factors affecting the probabilities. Happily, such a calculation is not too complicated. We have

$$P(C_3|M_5) = \frac{P(C_3)P(M_5|C_3)}{P(M_5)} = \frac{\left(\frac{1}{4}\right)(1)}{(1)\left(\frac{1}{4}\right) + \left(\frac{1}{2}\right)\left(\frac{3}{8}\right)} = \frac{4}{7}.$$

And since the probabilities of the two remaining doors must now sum to 1, we have the probability vector

$$\left(0, 0, \frac{4}{7}, \frac{3}{7}, 0\right).$$

Both probabilities went up, as predicted.

And if Monty had opened door three instead? Monty's failure to open door five when it seemed he had the freedom to do so will cause its probability to go up. On the other hand, by erasing the prior history of the game, Monty makes the probability of door four go down. The calculation looks like this:

$$P(C_5|M_3) = \frac{P(C_5)P(M_3|C_5)}{P(M_3)} = \frac{\left(\frac{3}{8}\right)(1)}{(1)\left(\frac{3}{8}\right) + \left(\frac{1}{2}\right)\left(\frac{3}{8}\right)} = \frac{2}{3},$$

leading to the probability vector

$$\left(0, 0, 0, \frac{1}{3}, \frac{2}{3}\right).$$

The probability of door four went down to $\frac{1}{3}$, reflecting the number of doors in play when we switched to it for the first time.

Intuition, however, is not a proof. It is time to get back to business.

4.6. Updating the Probabilities

If simple arguments are inadequate, we will turn instead to more complex ones.

Let us consider the moment in the game when only two doors remain in play. If we have been following SLM, then one of these doors will have probability $\frac{1}{n}$ and the other will have probability $\frac{n-1}{n}$. Outperforming SLM requires that when two doors remain, one of them must have a probability smaller than $\frac{1}{n}$. This will force the other to have probability greater than $\frac{n-1}{n}$.

Our proof that SLM is uniquely optimal will come in two parts. In this section we derive formulas that tell us how to update the probabilities for the surviving doors each time Monty reveals a goat. In the next section we will use these formulas to establish the impossibility either of obtaining a door of probability smaller than $\frac{1}{n}$ or of obtaining a door of probability equal to $\frac{1}{n}$ by any strategy other than SLM.

Let us now suppose that we are partway through the game and that k doors remain, with $3 \leq k \leq n$. As before, we denote by C_i the event that the car is behind door i and by M_j the event that Monty opens door j. We also define S_ℓ to denote the event that you select door ℓ. We can write

$$P(M_j | C_i \cap S_\ell) = \begin{cases} 0, & \text{if } j = i \text{ or } j = \ell; \\ \frac{1}{k-1}, & \text{if } i = \ell \text{ and } j \neq \ell; \\ \frac{1}{k-2} & \text{if } i \neq \ell, \, j \neq \ell \text{ and } j \neq i. \end{cases} \tag{4.1}$$

It follows from the definition of conditional probability that

$$P(M_j | C_i \cap S_\ell) = \frac{P(C_i \cap M_j | S_\ell)}{P(C_i | S_\ell)}. \tag{4.2}$$

Note, however, that C_i and S_ℓ are independent. Since choosing a door provides no basis for altering the probability of the door, we conclude that $P(C_i | S_\ell) = P(C_i)$. Therefore, (4.2) becomes

$$P(C_i) P(M_j | C_i \cap S_\ell) = P(C_i \cap M_j | S_\ell).$$

Combining this with (4.1) gives us

$$P(C_i \cap M_j | S_\ell) = \begin{cases} 0, & \text{if } j = i \text{ or } j = \ell; \\ \frac{1}{k-1} P(C_i), & \text{if } i = \ell \text{ and } j \neq \ell; \\ \frac{1}{k-2} P(C_i), & \text{if } i \neq \ell, \, j \neq \ell \text{ and } j \neq i. \end{cases} \tag{4.3}$$

Using the equation

$$P(C_i \cap M_j | S_\ell) = P(C_i | M_j \cap S_\ell) P(M_j | S_\ell),$$

(which again follows from the definition of conditional probability), we can recast equation (4.3) as

$$P(C_i | M_j \cap S_\ell) = \left(\frac{1}{P(M_j | S_\ell)} \right) \begin{cases} 0, & \text{if } i = j; \\ \frac{1}{k-1} P(C_i), & \text{if } i = \ell; \\ \frac{1}{k-2} P(C_i), & \text{if } i \neq \ell \text{ and } i \neq j. \end{cases} \tag{4.4}$$

Note that since it is impossible for Monty to open the same door you choose, in (4.4) we have implicitly assumed $j \neq \ell$. If we throw out the case where $i = j$, then equation (4.4) is the complete list of updated probabilities for the $k - 1$ doors remaining after Monty opens a door. This completes the first part of the proof.

4.7. SLM Is Uniquely Optimal

To complete the proof that SLM is uniquely optimal, we use equation (4.4) to derive certain results about the ratios of the probabilities of the remaining doors at any stage of the game.

Assume that k doors remain in play. Further suppose that you now choose door ℓ and Monty subsequently opens door $j \neq \ell$. It is a consequence of equation (4.4) that for $h \neq i$ we have

$$\frac{P(C_i | M_j \cap S_\ell)}{P(C_h | M_j \cap S_\ell)} = \frac{P(C_i)}{P(C_h)} \begin{cases} \frac{k-2}{k-1}, & \text{if } i = \ell; \\ \frac{k-1}{k-2}, & \text{if } h = \ell; \\ 1, & \text{if } h \neq \ell \text{ and } i \neq \ell. \end{cases}$$

We have already shown that when following SLM beginning with door 1, at the stage with k doors remaining, $P(C_1) = \frac{1}{n}$ and $P(C_i) = \frac{n-1}{n(k-1)}$ for $i \neq 1$. It follows that $\frac{P(C_1)}{P(C_i)} = \frac{k-1}{n-1}$.

Furthermore, we claim that with k doors remaining, we have

$$\frac{P(C_i)}{P(C_h)} \geq \frac{k-1}{n-1} \tag{4.5}$$

for any pair of doors remaining in play. We prove this by induction on the number of doors that have been opened.

At the beginning of the game $k = n$ and all doors have probability $\frac{1}{n}$. Since this is in accord with inequality (4.5), we see that the base case is satisfied.

Now suppose k doors remain unopened and all pairs of probabilities satisfy inequality (4.5). If you now choose door ℓ and Monty opens door j, the updated ratios are

$$\frac{P(C_\ell | M_j \cap S_\ell)}{P(C_h | M_j \cap S_\ell)} = \frac{k-2}{k-1}\left(\frac{P(C_\ell)}{P(C_h)}\right) \quad \text{and} \tag{4.6}$$

$$\frac{P(C_i | M_j \cap S_\ell)}{P(C_\ell | M_j \cap S_\ell)} = \frac{k-1}{k-2}\left(\frac{P(C_i)}{P(C_\ell)}\right) \tag{4.7}$$

if one of i or h is equal to ℓ. Since $\frac{P(C_\ell)}{P(C_h)}, \frac{P(C_i)}{P(C_\ell)} \geq \frac{k-1}{n-1}$, it is readily seen that these updated ratios satisfy the bound of inequality (4.5).

If we have $i, h \neq \ell$, then

$$\frac{P(C_i | M_j \cap S_\ell)}{P(C_h | M_j \cap S_\ell)} = \frac{P(C_i)}{P(C_h)} \geq \frac{k-1}{n-1},$$

and again we see that inequality (4.5) is satisfied. This completes the induction.

Now suppose that $k = 2$ and that door i and door h are the only ones remaining. Our previous result implies that

$$\frac{P(C_i)}{P(C_h)} = \frac{P(C_i)}{1 - P(C_i)} \geq \frac{1}{n-1}.$$

This implies that $P(C_i) \geq \frac{1}{n}$. Equivalently, $P(C_i) \leq \frac{n-1}{n}$. This establishes the optimality of SLM.

It remains only to establish that SLM is uniquely optimal. To do this, note that our previous calculations show that it is impossible, at any stage of the game, for a given door i to satisfy $P(C_i) < \frac{1}{n}$. Such a probability would imply that at least one of the other $k - 1$ unopened doors must have probability

$$P(C_h) > \frac{1 - \frac{1}{n}}{k - 1} = \frac{n - 1}{n(k - 1)}.$$

But then $\frac{P(C_i)}{P(C_h)} < \frac{k-1}{n-1}$, contradicting inequality (4.5).

Finally, note that in inequality (4.5) we obtain equality only if (a) our current door choice is placed in the numerator (as in equation (4.6)) and (b) at the previous stage the ratio of the probabilities for the same two doors was $\frac{k-1}{n-1}$. But (b) is possible only if door ℓ has been our choice all along. It follows that only by following SLM can the optimal ratio be obtained, and the proof is complete.

4.8. Monty Meets Shannon

Let us pause to consider the SLM strategy from a new angle.

Every time Monty opens a door he is giving us information concerning the location of the car. The optimal strategy will be the one that forces Monty to give us the maximum amount of information at each stage of the game. This raises a question: how ought we to quantify the amount of information we receive from Monty?

In discussing the amount of information received upon learning that a particular event has taken place, we are presupposing a situation in which there was prior uncertainty about what was going to happen. If we are certain that X will occur, then no information is received upon learning that X did in fact occur. Mathematically, we might imagine a collection of possible occurrences each with some probability attached to it. In some way the quantity of information received ought to be connected to the probability of the event. But how?

Intuitively there ought to be an inverse correlation between probability and information. That is, you receive more information from learning that a low-probability event has occurred than you do from learning that a high-probability event has occurred. When it is nearly certain that X will occur, you do not learn very much upon hearing that X has in fact occurred. If you hear instead that Y, a highly unlikely alternative to X, has occurred, then you have learned something far more informative.

The first person to develop these vague notions into a rigorous and useful theory was Claude Shannon in 1948. His interests lay in certain problems involving the efficient coding and transmission of messages through a communications channel such as a telegraph wire. He defined the "self-information" of an event E in a probability space by the formula

$$I(E) = -\log_2 P(E).$$

By defining things in this way he captured the intuitive idea that low probability corresponds to high information. His definition also has other advantages. For one, imagine that E_1 and E_2 are independent events. Then the probability that both occur is obtained by multiplying the probabilities of each one individually. To compute the quantity of information received when it is learned that both events occurred, note that

$$I(E_1 \cap E_2) = -\log_2[P(E_1)P(E_2)]$$
$$= -(\log_2 P(E_1) + \log_2 P(E_2)) = I(E_1) + I(E_2).$$

The information received from learning that each of two independent events occurred is the sum of the information contained in each event. Very nice.

It is not my purpose here to belabor the mathematical formalism developed by Shannon. I mention it simply to illustrate the connection between probability and information. It is a recurring theme of this book that you can teach an entire course in probability theory using nothing more than variations on the Monty Hall problem. Thinking in Shannon's terms provides a convenient way of viewing the SLM strategy.

Extracting the maximum quantity of information from Monty's actions requires forcing him to open high-probability doors. That is, the more likely we think a door is to contain the car, the more information we receive when Monty shows us that it is empty. In playing the progressive version of the game, we know that our initial choice is very likely to be incorrect. If we switch from that door at some point prior to the end of the game, then we are giving Monty the possibility of opening a low-probability door. That means that by switching we risk not receiving from Monty the maximal quantity of information he is prepared to give us. Such a strategy cannot be optimal.

4.9. Switching Every Time

Proving the optimality of SLM hardly exhausts the intrigue of the progressive Monty Hall problem.

It has been my experience in presenting this problem to my classes that the strategy of switching doors at every opportunity is invariably popular with students. This discussion typically comes after a long struggle to persuade them of the benefits of switching in the classical version. The take-home message seems to be that switching is a very good thing indeed and therefore ought to be done as frequently as possible.

In light of this, let us see what can be said on behalf of various non-optimal switching strategies. Our first solution had the advantage of using only the basic probabilistic machinery we have developed to this point, but such arguments quickly become unwieldy. We will therefore try a different approach to the problem, based on recurrence relations. First we analyze the strategy of switching every time. In the remainder of the chapter we will ponder other switching strategies.

We assume that every time Monty reveals a goat we select randomly from among the unopened doors different from our current choice. Denote by a_n the probability of winning with this strategy. Our analysis now splits into two cases, depending on whether our initial choice is correct or incorrect. Let b_n denote the probability of winning if we begin with n doors and our initial choice conceals a goat, and let c_n denote the probability of winning if we begin with n doors and our initial choice conceals the car. Since there is one car and $n - 1$ goats, we can write

$$a_n = \left(\frac{n-1}{n}\right) b_n + \left(\frac{1}{n}\right) c_n. \tag{4.8}$$

This can be seen as a variation on the law of total probability.

Suppose our initial choice conceals a goat. Monty now opens some other door and reveals a different goat. This time, among the remaining $n - 2$ doors different from our current choice, there is one that conceals the car and $n - 3$ that conceal goats. Thus, there is a probability of $\frac{1}{n-2}$ that we will switch to the car and a probability of $\frac{n-3}{n-2}$ that we will switch to a goat. Consequently, we can write

$$b_n = \left(\frac{n-3}{n-2}\right) b_{n-1} + \left(\frac{1}{n-2}\right) c_{n-1}. \tag{4.9}$$

Alternatively, suppose our current choice conceals the car. Again Monty reveals a goat. In this case, each of the remaining $n - 2$ doors will conceal a goat. It follows from our assumptions that we will definitely switch to a goat, and therefore

$$c_n = b_{n-1}. \tag{4.10}$$

We can use equation (4.10) to eliminate the c_i terms in equations (4.8) and (4.9). We obtain

$$a_n = \left(\frac{n-1}{n}\right) b_n + \left(\frac{1}{n}\right) b_{n-1} \tag{4.11}$$

and

$$b_n = \left(\frac{n-3}{n-2}\right) b_{n-1} + \left(\frac{1}{n-2}\right) b_{n-2}. \tag{4.12}$$

Note that $b_2 = 0$ and $b_3 = 1$.

Repeated applications of equation (4.12) lead to

$$b_{n+2} = \left(\frac{n-1}{n}\right) b_{n+1} + \left(\frac{1}{n}\right) b_n$$

$$= \left(\frac{n-1}{n}\right) \left(\frac{n-2}{n-1} b_n + \frac{1}{n-1} b_{n-1}\right) + \left(\frac{1}{n}\right) b_n$$

$$= \left(\frac{n-1}{n}\right) b_n + \left(\frac{1}{n}\right) b_{n-1}.$$

Since this last expression appears as the right-hand side of (4.11), we see that $b_{n+2} = a_n$. Rewrite (4.12) with $n + 2$ in place of n. This substitution gives us the following recurrence:

$$a_n = \left(\frac{n-1}{n}\right) a_{n-1} + \left(\frac{1}{n}\right) a_{n-2}, \tag{4.13}$$

with $a_0 = 0$ and $a_1 = 1$.

To solve (4.13), we rewrite it as

$$a_n - a_{n-1} = -\frac{1}{n}(a_{n-1} - a_{n-2}).$$

Since $a_1 - a_0 = 1$, we have

$$a_n - a_{n-1} = -\frac{(-1)^n}{n!},$$

which implies that the probability of winning is

$$a_n = a_n - a_0 = \sum_{i=1}^{n}(a_i - a_{i-1}) = -\sum_{i=1}^{n} \frac{(-1)^i}{i!}.$$

It follows that

$$\lim_{n \to \infty} a_n = -\sum_{i=1}^{\infty} \frac{(-1)^i}{i!} = 1 - \sum_{i=0}^{\infty} \frac{(-1)^i}{i!} = 1 - \frac{1}{e},$$

recognizing that the infinite series is that for e^x with $x = -1$. The number of doors n does not need to be very large before the probability stabilizes at 0.632.

It is an interesting fact that the quantity $1 - \frac{1}{e}$ comes up in a large number of problems in elementary probability. Have a look at [55] for some examples.

4.10. Other Strategies

What can be said on behalf of other switching strategies? We note first that equation (4.8), though formulated with the "switch every time" strategy in mind, is actually valid for any strategy. More specifically, let S be a given strategy and denote by a_n the probability of winning with S at the moment when n doors remain in play. We now make our door choice. If b_n denotes the probability of winning with S given that our current choice conceals a goat, and c_n denotes the probability of winning given that our current choice conceals the car, then a_n, b_n, and c_n are related via equation (4.8).

Furthermore, since it is assumed the doors are identical, and therefore that their numbering is arbitrary, we only need to consider strategies that call for switching at specific moments during the game. Strategies such as "switch if Monty opens an even door, but stick otherwise" cannot be optimal, and we will not consider them.

Let us assume, then, that our strategy calls for us to switch doors a total of k times with $k \leq n - 2$. We will assume that we choose our new door randomly from the available options each time we switch. Denote by $\{m_i\}_{i=1}^{k}$ the number of doors remaining when we make the $(k - i + 1)$-st switch. We have $3 \leq m_1 < m_2 < \ldots < m_k \leq n$.

For any integer j, the manner in which the probabilities b_j and c_j are related to b_{j-1} and c_{j-1} will depend on whether or not we switch at the moment when j doors remain. If we switch, then the probabilities are related in the manner described by equations (4.9) and (4.10). If we do not switch, then the probabilities do not change. To see this, note that our probability of winning by sticking with our present door is equal to the probability that it conceals the car. A straightforward argument using Bayes' theorem now shows that this probability cannot change so long as we maintain this door as our selection. Consequently, our probability of winning can change only at those moments of the game when we decide to switch doors.

It follows that we have

$$b_{m_i} = \left(\frac{m_i - 3}{m_i - 2} \right) b_{m_{i-1}} + \left(\frac{1}{m_i - 2} \right) c_{m_{i-1}} \text{ and } c_{m_i} = b_{m_{i-1}}, \qquad (4.14)$$

for all $1 \leq i \leq k$. For any subscript $j \neq m_i$ for any i, we have $b_j = b_{j-1}$ and $c_j = c_{j-1}$.

To simplify the notation, we define $\beta_i = b_{m_i}$ and $q_i = m_i - 2$. This leads to

$$\beta_i = \left(\frac{q_i - 1}{q_i} \right) \beta_{i-1} + \left(\frac{1}{q_i} \right) \beta_{i-2}, \qquad (4.15)$$

for $i = 1, 2, \ldots, k$. Note that we have the initial conditions $\beta_{-1} = 1$ and $\beta_0 = 0$. The probability of winning given there are n doors and a given set of k door changes is thus

$$a_n = \left(\frac{n - 1}{n} \right) \beta_k + \left(\frac{1}{n} \right) \beta_{k-1}. \qquad (4.16)$$

To solve (4.16), set $\gamma_i = \beta_i - \beta_{i-1}$. Subtracting β_{i-1} from both sides of equation (4.15) then leads to

$$\gamma_i = \frac{-\gamma_{i-1}}{q_i},$$

with $\gamma_0 = -1$. Thus $\gamma_1 = 1/q_1, \gamma_2 = -1/(q_1 q_2), \gamma_3 = 1/(q_1 q_2 q_3)$, and in general

$$\gamma_j = \frac{(-1)^{j+1}}{\prod_{i=1}^{j} q_i}.$$

Since $\beta_i = \gamma_i + \beta_{i-1}$,

$$\beta_1 = \frac{1}{q_1}, \quad \beta_2 = \frac{1}{q_1} - \frac{1}{q_1 q_2}, \quad \beta_3 = \frac{1}{q_1} - \frac{1}{q_1 q_2} + \frac{1}{q_1 q_2 q_3},$$

and in general

$$\beta_j = \frac{1}{q_1} - \frac{1}{q_1 q_2} + \frac{1}{q_1 q_2 q_3} - \ldots + \frac{(-1)^{j+1}}{\prod_{i=1}^{j} q_i}. \tag{4.17}$$

Finally, substitution back in (4.16) gives us

$$a_n = \frac{1}{q_1} - \frac{1}{q_1 q_2} + \frac{1}{q_1 q_2 q_3} - \ldots + \frac{(-1)^{k+1}}{\prod_{i=1}^{k} q_i} - \frac{(-1)^{k+1}}{n \prod_{i=1}^{k} q_i}. \tag{4.18}$$

This equation allows us to answer the following question: Suppose we are playing the game with n doors and are absolutely determined to switch some number k times. What is our optimal strategy in this situation? Let us begin by analyzing a few simple cases.

If we never change doors, then $k = 0$, $\beta_0 = 0$, and $\beta_{-1} = 1$. It follows that $a_n = \frac{1}{n}$. This makes sense. Our initial choice is correct with probability $\frac{1}{n}$, and this probability cannot change so long as it remains our choice. Note that as n increases, the probability of winning approaches 0.

If we change doors exactly once, then $k = 1$ and

$$a_n = \frac{1}{q_1} - \frac{1}{n q_1} = \frac{n-1}{n q_1}.$$

The probability of winning is maximized, given our constraints on q_1, by choosing $q_1 = 1$, which is equivalent to $m_1 = 3$. This corresponds to switching at the last possible moment. Again, this makes sense. If you are going to switch only one time, you should do so after Monty opens his final door, as we have already shown. This strategy gives a probability of winning $a_n = \frac{n-1}{n}$, which approaches 1 as n increases.

If we change doors twice, $k = 2$ and

$$a_n = \frac{1}{q_1} - \frac{1}{q_1 q_2} + \frac{1}{n q_1 q_2} = \frac{n(q_2 - 1) + 1}{n q_1 q_2}.$$

This expression is maximized by choosing q_1 to be as small as possible, which implies that $q_1 = 1$. After making that substitution, we could employ some elementary calculus to establish that the right-hand side of (4.10) is an increasing function of q_2 and therefore is maximized by setting $q_2 = n - 2$. This corresponds to making our first switch as early in the game as possible, then waiting until the last possible moment to make the second switch. This strategy wins with probability

$$a_n = \frac{n^2 - 3n - 1}{n^2 - 2n}.$$

Unsurprisingly, our best strategy when switching twice does not win as frequently as our best strategy when switching once.

Finally, if we change doors three times, then $k = 3$ and

$$a_n = \frac{1}{q_1} - \frac{1}{q_1 q_2} + \frac{1}{q_1 q_2 q_3} - \frac{1}{n q_1 q_2 q_3}.$$

Since $1/q_1$ is a common factor to all terms, we see that a_n is maximized by minimizing q_1. That is, we set $q_1 = 1$. Determining the appropriate values for q_2 and q_3 is trickier. The value of a_n increases with q_2 (suggesting that q_2 should be maximized) but decreases with q_3 (suggesting that q_3 should be minimized). We must balance these considerations with the fact that $q_2 < q_3$. This is accomplished by setting $q_3 = q_2 + 1$. With this substitution, it is straightforward to show that we should take $q_2 = n - 3$ and $q_3 = n - 2$. This corresponds to making the first two switches immediately and then waiting until the end to make the third. This wins with probability

$$a_n = \frac{n^3 - 6n^2 + 9n - 1}{n(n - 2)(n - 3)}.$$

Based on these examples, you might suspect that if you are determined to switch exactly $k > 2$ times, your best strategy is to make your first $k - 1$ switches as soon as possible and then wait until the last possible minute to make your final switch. That suspicion is correct, as the following theorem shows:

Theorem 1 Let S be a strategy for the progressive Monty Hall problem that calls for you to switch exactly k times, with $2 \leq k \leq n - 2$. Then the probability of winning with S is maximized by making switches $1, 2, 3, \ldots, k - 1$ when there are respectively $n - 1, n - 2, \ldots, n - k + 1$ doors remaining, and making the k-th switch when only two doors remain.

Proof 1 The quantity to be maximized is a_n from equation (4.18), subject to the constraints that $q_i \in \mathbb{Z}$ for all i and

$$1 \leq q_1 < q_2 < \ldots < q_k \leq n - 2.$$

Equation (4.18) can be rewritten as

$$a_n = \frac{1}{q_1}\left(1 - \frac{1}{q_2}\left(1 - \frac{1}{q_3} + \frac{1}{q_3 q_4} + \ldots + \frac{(-1)^{k+1}}{\prod_{i=3}^{k} q_i} - \frac{(-1)^{k+1}}{n\prod_{i=3}^{k} q_i}\right)\right).$$

It is clear that a_n is maximized by setting $q_1 = 1$. This corresponds to making your final switch at the last possible moment.

To simplify the notation, set

$$\gamma_k = 1 - \frac{1}{q_3} + \frac{1}{q_3 q_4} + \ldots + \frac{(-1)^{k+1}}{\prod_{i=3}^{k} q_i} - \frac{(-1)^{k+1}}{n\prod_{i=3}^{k} q_i}.$$

Maximizing a_n now requires that we minimize γ_k/q_2.

To do this, first note that γ_k is an alternating series whose terms are strictly decreasing in magnitude. It follows that $\gamma_k \geq 1 - (1/q_3)$. Since $q_3 > 1$, this implies that $\gamma_k > 0$.

Next, view the function $f(q_1, q_2, \ldots, q_k) = \gamma_k/q_2$ as a function from $\mathbb{R}^k \to \mathbb{R}$; that is, allow the q_i's to be real-valued instead of integer-valued. Then we can evaluate the partial derivative with respect to q_2, which gives us

$$\frac{\partial f}{\partial q_2} = \frac{\gamma_k}{(q_2)^2} > 0,$$

for all values of q_2. It follows that γ_k/q_2 will be minimized when q_2 is maximized. With our constraints, that means setting $q_2 = n - k$.

But this, in turn, forces us to set $q_i = n - k + (i - 2)$ for $3 \leq i \leq k$. This corresponds to the strategy laid out in the theorem, and the proof is complete. □

Table 4.2: Probabilities of winning given k switches and n doors (to four digits accuracy)

$k\backslash n$	3	5	10	20	50	100	200	500
1	0.6667	0.8000	0.9000	0.9500	0.9800	0.9900	0.9950	0.9980
2		0.7333	0.8875	0.9472	0.9896	0.9899	0.9950	0.9980
3		0.6333	0.8732	0.9443	0.9792	0.9898	0.9949	0.9980
4			0.8545	0.9410	0.9787	0.9897	0.9949	0.9980
5			0.8291	0.9373	0.9783	0.9896	0.9949	0.9980
8			0.6321	0.9226	0.9767	0.9892	0.9948	0.9980
18				0.6321	0.9697	0.9880	0.9945	0.9979
48					0.6321	0.9811	0.9935	0.9978
98						0.6321	0.9903	0.9975
198							0.6321	0.9967
498								0.6321

4.11. Revisiting the Optimality of SLM

Our exertions with the recurrence relations permit an alternative proof of the optimality of SLM. Let us close the technical portion of this chapter by having a look.

Theorem 2   SLM is the uniquely optimal strategy for the progressive Monty Hall problem.

Proof 2   We begin by assuming that we change doors exactly k times, where $k > 1$. We have already shown that the $k = 1$ case (one switch at the end) is superior to the best $k = 2$ or $k = 3$ cases. We worked out the optimal strategy for $k > 3$ switches in the previous section. Using this strategy, the probability of winning is

$$a_n = 1 - \frac{1}{(n-k)} + \frac{1}{(n-k)(n-k+1)} - \ldots +$$

$$\frac{(-1)^k}{(n-k)(n-k+1)\ldots(n-2)} - \frac{(-1)^k)}{n(n-k)(n-k+1)\ldots(n-2)}.$$

This is an alternating series where each term is strictly smaller in magnitude than its predecessor, and so has an upper bound

$$a_n < 1 - \frac{1}{(n-k)} + \frac{1}{(n-k)(n-k+1)} = 1 - \frac{1}{(n-k+1)} < 1 - \frac{1}{n}.$$

Thus, the optimum when changing doors more than once is an inferior strategy to changing doors at the last possible moment, completing the proof.   □

Table 4.2 shows the probability of winning given various assumptions about k and n.

There is more that can be said regarding the progressive Monty Hall problem. Our equation (4.18) appears also in the paper by Paradis, Viader, and Bibiloni [75]. Curiously, their derivation is quite different from the one presented here. They use only elementary probability theory.

Writing in [21], Engel and Venetoulias provide an alternative method for deriving the optimal strategies. Their method requires a level of probabilistic sophistication well beyond the level of this book, but it is correspondingly more powerful as well. They also go on to consider certain variations of the basic progressive game. Fascinating stuff, but it will have to wait for a different book.

5

Miscellaneous Monty

We have trodden a long and winding road to reach this point. We have navigated the rapids of the classical problem and some of its most natural variations. We confronted the full horror of the progressive version and emerged stronger for the experience. Along the way we have illuminated much of the world of probability theory and its offshoots. Yet for all of that, there remain certain variations on and aspects of the Monty Hall problem that have not fit comfortably into the preceding chapters. Our purpose now is to tie up some of those loose ends.

5.1. Benevolent and Malevolent Monty

Let me note at the outset that the variations considered here and in the next section are drawn from the Wikipedia article on the Monty Hall problem.

It has been a recurring theme that your optimal strategy in the Monty Hall problem depends in part on the precise manner in which Monty makes his decisions regarding which door to open. Let us catalog a few possibilities we have not previously considered.

Suppose Monty only offers the option of switching when your initial choice is the winning door. In this case switching will always yield a goat. By contrast, if Monty only offers the option of switching when you have chosen incorrectly, then switching will always yield the car.

Table 5.1: Probabilities of winning by switching given different host behaviors

Host	Bad	Middle	Best
Benevolent Monty	0	1/3	2/3
Random Monty	1/3	1/3	1/3
Malevolent Monty	2/3	1/3	0

Of course, if Monty always gives you the option of switching and only opens goat-concealing doors (and chooses randomly from among the goat-concealing doors when he has a choice of doors to open), then you win with probability $\frac{2}{3}$ by switching. This corresponds to the classical version of the problem.

The possibility of Monty choosing his door randomly was the subject of Chapter 3. In this case we stipulate that you lose if Monty reveals the car, which happens with probability $\frac{1}{3}$. As we have seen, you win with probability $\frac{1}{2}$ by switching in the cases where Monty reveals a goat.

Interestingly, the analysis for Random Monty remains unchanged even if we allow Monty the option of opening the door you selected for your initial choice. Assuming that Monty chooses randomly and that you lose if he reveals the car, you will still lose immediately one-third of the time, and win with probability $\frac{1}{2}$ by switching in the remaining cases.

Building from this foundation, we now provide an interesting way of looking at things. Let us suppose there are three different prizes, ranked unambiguously as bad, middle, and good. Let us further distinguish three possible behaviors from Monty. Benevolent Monty always reveals the worst remaining prize after you make your initial choice (never opening the door you chose initially). Random Monty chooses randomly from among the unchosen doors, while Malevolent Monty always reveals the best remaining prize. If you follow a strategy of always switching, your probabilities of success break down as in Table 5.1.

By now we are old pros at working out such probabilities, but let us pause to consider them nevertheless. The key to understanding Benevolent Monty is that he will never reveal the best prize. This scenario is therefore equivalent to the classical problem, with the best prize playing the role of the car and the other two prizes playing the roles of the goats. Likewise, Malevolent Monty will never reveal the bad prize, placing it in the role of the car. But Random Monty has an equal probability of opening any of the three doors. That makes this scenario comparable to the first version of Chapter 3.

Thus, deriving an advantage from switching requires a knowledge of the host's behavior and the nature of the prizes being offered. If Monty opens his doors without any restrictions regarding their contents, then we derive no useful information about the location of the best remaining prize from Monty's actions. But if his door opening is biased in some direction that is known to us, then we can improve our chances.

5.2. Two Players

We now consider a two-player version of the game:

> **Version Seven:** *As usual, we are presented with three doors. This time, however, there is a second player in the game. Player one chooses a door, and then player two chooses a different door. If both have chosen goats, then Monty eliminates one of the players at random. If one has chosen the car, then the other player is eliminated. The surviving player knows the other has been eliminated, but does not know the reason for the elimination. After eliminating a player, Monty then opens that player's door and gives the surviving player the options of switching or sticking. What should the player do?*

The scenario can be approached from a variety of directions. The simplest is to note that switching now wins only if both players selected goats. The probability of selecting two goats in two door choices is equal to the probability of choosing the car in one door choice. That is to say, the probability is $\frac{1}{3}$. We therefore have the rare example of a Monty Hall variation in which the best strategy is to stick.

An alternative approach involves dividing the possibilities into the following three, equally likely, cases:

- Player one selects the car. Monty must eliminate player two. Switching loses.

- Player two selects the car. Monty must eliminate player one. Switching loses.

- Neither player picks the car. One player is eliminated at random. Switching wins.

We arrive once more at the conclusion that switching is the wrong approach in this case.

The really interesting part of this example comes when we compare it with the classical version. From the perspective of the surviving player, things seem identical to the classical version. In both cases the player chooses a door, sees a goat revealed under circumstances where Monty never reveals the car, and is then presented with his options. Why, then, the different strategies in the two cases? What has changed?

The answer lies in the extra information gleaned by the surviving player from the very fact that he survived. For simplicity, assume that player one is the surviving player. Let us assume that his current choice conceals a goat, and therefore that he will win by switching. For this to be the case, we must assume not only that both players have chosen goats but also that player two was then chosen randomly for elimination. If we denote by G the event that player one's current choice conceals a goat, by B the probability that both players chose goats, and by S the probability that player one survives given that both players selected goats, then we have

$$P(G) = P(B)P(S|B) = \frac{2}{3} \times \frac{1}{2} = \frac{1}{3},$$

precisely as before.

You see, in the classical version there is no chance that the player's door will be eliminated. Consequently, the player receives no information from his continued presence in the game. But in the two-player version each player faces a nonzero probability of being eliminated. Not being eliminated therefore constitutes information that must be incorporated into any further decisions.

The general principle for the three-door versions we have considered is that if a particular door has a 0 probability of being opened, then the probability of that door does not change as the result of Monty's actions. In the classical version it is the player's initial choice that cannot be opened. Consequently, it retains its $\frac{1}{3}$ probability even after Monty does his thing. In the two-player version it is the unchosen door that cannot be eliminated. Thus, this is the door that retains its $\frac{1}{3}$ probability of being correct. If we were to consider a three-player version, in which each player chooses a different door and one of the goats is eliminated at random, then we would have a situation very similar to version two of the problem from Chapter 3. In this case there would be no advantage to be obtained from switching.

5.3. Two Hosts

Here's an amusing exercise I found in [76]:

> *Version Eight: As before, we are confronted with three identical doors, one concealing a car, the other two concealing goats. We initially choose door one, and Monty then opens door three. This time we know that there are two different hosts who preside over the show, with a coin flip deciding who hosts the show on a given night. The two hosts do not make their decisions in the same way. Coin-Toss Monty chooses his door randomly when your initial choice conceals the car. Three-Obsessed Monty always opens door three when he has the option of doing so. Under these circumstances, is there an advantage to be gained from switching to door two?*

We will continue to denote by M_i the event in which Monty opens door i, and by C_i the event in which the car is behind door i. To simplify the notation, we will write

$$p = P(C_1|M_3) \text{ and } q = P(M_3|C_1).$$

It is a consequence of the work we did in section 3.8 that, so long as the three doors are equaly likely at the start and Monty is guaranteed to reveal a goat, we have $p = \frac{1}{1+q}$.

Notice that for Coin-Toss Monty we have $q = \frac{1}{2}$ and for Three-Obsessed Monty we have $q = 1$. These values correspond to $p = \frac{2}{3}$ and $p = \frac{1}{2}$ respectively, as shown in our previous work. We might be inclined to argue that since we know each host is chosen with probability $\frac{1}{2}$, then our probability of winning by switching in this scenario is simply the average of $\frac{2}{3}$ and $\frac{1}{2}$, which is $\frac{7}{12}$. This seems reasonable. We know that if the host chooses his door randomly when given a choice, we gain a big advantage from switching. If the host is sure to open door three when that option is viable, then switching provides no advantage. Since there is an equal probability of getting either host, it seems fair to conclude that on average there will be a small advantage to be gained from switching, precisely as this argument suggests.

If you find this argument appealing, then I fear you have not fully assimilated the lessons of the previous chapters. It is true that when the game began we assigned an equal probability to obtaining either host, because the problem stipulated that we should do so. But after Monty takes some action, opening door three in this case, we have new information regarding the host we have drawn.

We could run things through Bayes' theorem to determine the probabilities that we have drawn each of the hosts in light of the fact that we have seen the host open door three, but since this situation is readily handled with the proportionality principle from Chapter 3, we may as well go that route.

Keep in mind we are assuming that you initially select door one.

Coin-Toss Monty opens door three with certainty when the car is behind door two, and with probability $\frac{1}{2}$ when the car is behind door one. Since the car is equally likely to be behind any of the three doors, we see that Coin-Toss Monty opens door three with probability

$$\frac{1}{3}(1) + \frac{1}{2}\left(\frac{1}{3}\right) = \frac{1}{2}.$$

On the other hand, Three-Obsessed Monty opens door three with certainty when the car is behind door one or door two. This happens with probability $\frac{2}{3}$.

We conclude that we are $\frac{4}{3}$ more likely to see the host open door three if we have drawn Three-Obsessed Monty than if we have drawn Coin-Toss Monty. By the proportionality principle our updated probabilities must preserve this ratio. Thus, we conclude that we have drawn Coin-Toss Monty with probability $\frac{3}{7}$ and Three-Obsessed Monty with probability $\frac{4}{7}$.

Our probability of winning by switching will therefore be the following weighted average:

$$\left(\frac{4}{7}\right)\left(\frac{1}{2}\right)+\left(\frac{3}{7}\right)\left(\frac{2}{3}\right)=\frac{4}{7}.$$

5.4. Many Doors, Many Cars

Writing in [31], John Georges and Timothy Craine offer a nice collection of variations based on certain multiple-door scenarios. We shall devote this and the next two sections to a consideration of their offerings.

We begin with a warm-up exercise. As in the progressive version of the problem, we assume there are $n \geq 3$ doors concealing one car and $n-1$ goats. After we make our initial selection. Monty opens a door he knows to conceal a goat, careful to choose randomly from among the doors different from your choice that do not conceal the car. Monty now gives us the option of switching, at which point we receive whatever is behind our door. Should we switch?

After all of our recent exertions, this one should seem like a breath of fresh air. We established in Chapter 4 that the probability of our initial choice does not change when Monty opens a door. Consequently, by sticking we win with probability $\frac{1}{n}$. We will win by switching precisely when, first, our initial choice does not conceal the car (which happens with probability $\frac{n-1}{n}$) and, second, when we then switch to the car (which happens with probability $\frac{1}{n-2}$). It follows that by switching we win with probability $\frac{n-1}{n(n-2)}$. And since $\frac{n-1}{n-2} > 1$, we see there is an advantage to be gained from switching. The size of that advantage, alas, approaches zero as the number of doors approaches infinity.

That was nice. How about another? This time we still have n doors, but now there are $1 \leq j \leq n-2$ cars and $n-j$ goats. After making your initial choice, Monty opens one of the other doors at random. Should you switch?

Since there are j cars and n doors, we see that our initial choice conceals a car with probability $\frac{j}{n}$. This probability will not change regardless of what Monty reveals, and therefore represents the probability of obtaining a car by sticking.

The probability of winning by switching, however, is a bit more complicated. Let us denote by F_c and F_g the events that our first choice is a car or a goat respectively. Similarly, let S_c and S_g denote the events that our second choice is a car or a goat. If we denote by P_{switch} the probability of winning by switching, then we have the general formula

$$P_{\text{switch}} = P(F_g)P(S_c|F_g) + P(F_c)P(S_c|F_c).$$

It is clear that

$$P(F_g) = \frac{n-j}{n} \quad \text{and} \quad P(F_c) = \frac{j}{n}.$$

The trouble comes with the conditional probabilities. Since their values depend on whether Monty reveals a car or a goat, we will need to divide things into cases.

After Monty reveals a goat, there are $n - 2$ doors still in play that are different from your initial choice. Among those doors are j cars and $n - j - 1$ goats. It follows that

$$P(S_c|F_g) = \frac{j}{n-2} \text{ and } P(S_c|F_c) = \frac{j-1}{n-2}.$$

This leads to the conclusion that

$$P_{\text{switch}} = \frac{n-j}{n}\left(\frac{j}{n-2}\right) + \frac{j}{n}\left(\frac{j-1}{n-2}\right) = \frac{j}{n}\left(\frac{n-1}{n-2}\right).$$

Since $\frac{n-1}{n-2} > 1$, we see once again that we increase our chances of winning by switching doors.

The surprising part comes when Monty reveals a car. (Of course, for this scenario to be interesting, we will have to assume that there are at least two cars in play, so that $j \geq 2$.) For now we have

$$P(S_c|F_g) = \frac{j-1}{n-2} \text{ and } P(S_c|F_c) = \frac{j-2}{n-2}.$$

Consequently, we compute that

$$P_{\text{switch}} = \frac{n-j}{n}\left(\frac{j-1}{n-2}\right) + \frac{j}{n}\left(\frac{j-2}{n-2}\right).$$

The curious thing is that with a little elementary algebra we can show that this quantity is smaller than $\frac{j}{n}$, which was the probability of winning by sticking. It would seem that in this case you would do better to stick with your original choice.

Let us try a new wrinkle. We still have n doors, with j cars and $n - j$ goats. This time, however, Monty tells us that he will reveal a goat with probability p, and will reveal a car with probability $1 - p$. The catch is that we must make our decision to stick or switch before knowing which of these possibilities will come to pass. What should we do?

Of course, the probability of winning by sticking is still $\frac{j}{n}$.

We have already worked out P_{switch} for the cases where Monty reveals a goat and where Monty reveals a car. In the present scenario, there is uncertainty regarding which of these possibilities will actually occur. Consequently, the value of P_{switch} in this case will be the weighted average of these two possibilities, and we will obtain

$$P_{\text{switch}} = p\left[\frac{n-j}{n}\left(\frac{j}{n-2}\right) + \frac{j}{n}\left(\frac{j-1}{n-2}\right)\right]$$

$$+(1-p)\left[\frac{n-j}{n}\left(\frac{j-1}{n-2}\right) + \frac{j}{n}\left(\frac{j-2}{n-2}\right)\right].$$

It follows that there is an advantage to be gained from switching precisely when the expression above is greater than $\frac{j}{n}$. With a bit of algebra we can see that the expressions are equal when we have $p = \frac{n-j}{n}$. In other words, if the probability that Monty reveals a goat is equal to the probability that a randomly chosen door conceals a goat, then it makes no difference whether you stick or switch. If $p > \frac{n-j}{n}$, then there is an advantage to be gained from switching. This makes good, intuitive sense. Our previous calculations show that we should switch when Monty reveals a goat and should stick when Monty reveals a car. If Monty chooses a door randomly, then he will choose a goat with probability $\frac{n-j}{n}$. Seen that way, it is logical that if Monty reveals a goat with greater probability than $\frac{n-j}{n}$, then we should switch, and if he reveals a goat with smaller probability, then we should stick.

Georges and Craine consider one further variation on this basic scenario. It will make a fitting conclusion to all our hard work.

What happens if Monty opens several doors? We still have n doors with j cars and $n - j$ goats. You choose a door, which conceals a car with probability $\frac{j}{n}$. This time Monty opens m doors, revealing k cars and $m - k$ goats. To avoid trivial scenarios, we shall assume that $1 \leq m \leq n - 2$ and $0 \leq k \leq m$. As always, the probability of winning by sticking is $\frac{j}{n}$, and the probability of winning by switching is given by the formula

$$P_{\text{switch}} = P(F_g)P(S_c|F_g) + P(F_c)P(S_c|F_c).$$

Also as before, we still have that

$$P(F_g) = \frac{n-j}{n} \text{ and } P(F_c) = \frac{j}{n}.$$

The novelty here lies in the values of the conditional probabilities. After Monty opens m doors, there remain $n - m - 1$ doors different from your current choice to switch to. Furthermore, there are $j - k$ cars still in play. It follows that

$$P(S_c|F_g) = \frac{j-k}{n-m-1} \text{ and } P(S_c|F_c) = \frac{j-k-1}{n-m-1}.$$

Assembling everything leads to the finding that

$$P_{\text{switch}} = \frac{n-j}{n}\left(\frac{j-k}{n-m-1}\right) + \frac{j}{n}\left(\frac{j-k-1}{n-m-1}\right)$$

$$= \frac{nj - nk - j}{n(n-m-1)}.$$

That is hardly the prettiest fraction in the world, but it does have an interesting interpretation. When the game started the ratio of cars to doors was $\frac{j}{n}$. Among the m doors Monty opened we find k cars, for a ratio of $\frac{k}{m}$. If these ratios are the same, then some straightforward algebra tells us that the expression above is equal to $\frac{j}{n}$. In other words, your probability of winning by switching is equal to your probability of winning by sticking. Likewise, if $\frac{k}{m} < \frac{i}{n}$, it turns out there is an advantage to switching. And if $\frac{k}{m} > \frac{i}{n}$? Then you do better to stick.

In other words, if the ratio of cars to doors in the sample Monty opens is smaller than the ratio of cars to doors when the game begins, then there is an advantage to switching. If these ratios are the same, then it does not matter what you do. Otherwise, you do best to stick. A satisfying result.

5.5. Expectation

Georges and Craine go on to consider one further variation. It has sufficient novelty to deserve a number of its own:

> *Version Nine: Again we have n doors with $n \geq 3$. Suppose we have m types of prize with values $v_1, v_2, \ldots, v_m$. Assume that for $1 \leq i \leq m$ there are n_i prizes with value v_i. After we make our initial choice, Monty opens one of the unchosen doors at random without regard for what is behind it. Under what circumstances should we switch?*

This is fundamentally different from all of our previous versions of the problem, for winning is now a nebulous concept. To this point our goal has always been to maximize the probability of winning a car, and we asked whether or not switching would increase that probability. This time, however, we want to know whether switching is likely to result in our winning a prize of value greater than or equal to the prize behind our initial choice. The mathematical machinery we have developed thus far is not well suited to that question.

Certainly we could divide things into cases and go from there. For example, we might begin by assuming that our initial choice concealed a prize of value v_1 and Monty revealed a prize of value v_2, and then work out the probabilities of switching to a door of value v_i for all possible i. This gets cumbersome in a hurry. A better approach is to think about the long run. Imagine that we play the game a large number of times and switch every time. What can we say about the average value of the prizes we win in this way? Textbooks on probability refer to this average as the expected value of our winnings, and we devote the remainder of this section to a consideration of that concept.

Let us suppose you roll a fair six-sided die multiple times, keeping track of your results. In the long run we would expect each of the six numbers to appear with relative frequency $\frac{1}{6}$. The average of the numbers one through six is 3.5. If each of these six numbers appears with the same relative frequency in our long-run of dice throws, then we would expect the average of our throws to be roughly 3.5 as well.

This observation generalizes readily. If X can take on any of n values with equal probability, then the average obtained from a long run of trials ought to be very close to the average of the n values.

But what if the values X can take on do not occur with equal probability? As a simple example, suppose that X takes on the value a with probability $\frac{3}{4}$ and the value b with probability $\frac{1}{4}$. Then in a long run of trials we expect to get a roughly three-fourths of the time and to get b roughly one-fourth of the time. That is, if we perform the experiment n times then we will obtain a roughly $\left(\frac{3}{4}\right) n$ times and b roughly $\left(\frac{1}{4}\right) n$ times.

Let us denote our expected average by $E_{av}(X)$. Keep in mind that the average value is obtained by adding up the results of the individual trials and then dividing by the total number of trials (which is n in this case). We have

$$E_{av}(X) = \frac{\left(\frac{3}{4}\right) na + \left(\frac{1}{4}\right) nb}{n} = \frac{3a}{4} + \frac{b}{4}.$$

The average value of X in the long run is very likely to be the average of the individual values X might take on, weighted by their probabilities of occurring.

We need some jargon here. A quantity X that can take on any of finitely many numerical values will be referred to as a **discrete random variable**. The probabilities with which X takes on each of its possible values will be referred to as the **probability distribution** of X. The **expected value** of X, denoted by $E(X)$, will then be the average of the possible values of X, with each value weighted by its probability of occurring. In other words, if X takes on values $a_1, a_2, \ldots, a_k$ with probabilities $p_1, p_2, \ldots, p_k$ respectively, then we have

$$E(X) = p_1 a_1 + p_2 a_2 + \ldots + p_k a_k.$$

As an example, imagine that we toss three coins. Let X denote the number of heads that come up. Then X can take on the four values $0, 1, 2,$ or 3. That is, the three coins can give you either zero, one, two, or three heads. What is the probability distribution for X? We note that the probability of obtaining zero heads or three heads is simply $\frac{1}{8}$. The probability of obtaining either one or two heads is readily seen to be $\frac{3}{8}$. Finally, we can write

$$E(X) = \left(\frac{1}{8}\right) 0 + \left(\frac{3}{8}\right) 1 + \left(\frac{3}{8}\right) 2 + \left(\frac{1}{8}\right) 3 = \frac{3}{2}$$

If we repeatedly toss three coins in the air, we expect the average number of heads that come up to be 1.5.

In practice you should think of a random variable as a bookkeeping device. It allows you to ignore the extraneous details of the full sample space and to focus instead on some particular quantity of interest. In the example above, we were not so interested in the specific sequence of heads and tails that arose each time we flipped the coins. Instead we cared only about the number of heads. The probability distribution of X is then worked out from the appropriate sample space. Finally, the expected value should be thought of as the average value X will take on in a long run of trials.

5.6. Many Types of Prize

Let us consider a concrete formulation of version nine before tackling it in full generality. Suppose there are six doors. They conceal two cars worth $20,000 each, two motorcycles worth $10,000 each, and two new refrigerators worth $300 each. We make an initial choice, and then Monty opens some other door at random without regard for what is behind it. We are given the options of sticking or switching. What should we do?

Consider the sticking strategy. In choosing our initial door we have a $\frac{1}{3}$ probability of winning a prize worth $20,000, a $\frac{1}{3}$ probability of winning a prize worth $10,000, and a $\frac{1}{3}$ probability of winning a prize worth $300. We know these probabilities do not change when Monty opens a door. If we let X denote the amount of money we win by sticking with our initial choice, then the expected value of sticking is

$$E_{\text{stick}}(X) = \frac{1}{3}(20{,}000) + \frac{1}{3}(10{,}000) + \frac{1}{3}(300) \approx 10{,}100.$$

In other words, if we play the game a large number of times and stick every time, then our average winnings are very likely to be just over $10,000.

The expected value of switching will depend both on what Monty reveals and on what is behind our initial choice. Consequently, we shall have to divide things into cases. The good news is that since Monty opens his door with no regard for what is behind it, each of the remaining doors has an equal probability of containing any of the remaining prizes.

Now let us suppose that Monty reveals a car. If our initial choice also concealed a car, then the remaining four unopened, unchosen doors conceal two motorcycles and two refrigerators. We compute

$$E(X) = \frac{1}{2}(10{,}000) + \frac{1}{2}(300) = 5{,}150.$$

If our initial choice concealed a motorcycle, then we obtain

$$E(X) = \frac{1}{4}(20{,}000) + \frac{1}{4}(10{,}000) + \frac{1}{2}(300) = 7{,}650.$$

And if our initial choice concealed a refrigerator, then we obtain

$$E(X) = \frac{1}{4}(20{,}000) + \frac{1}{2}(10{,}000) + \frac{1}{4}(300) = 10{,}075.$$

Since our initial choice contains a car, motorcycle, or refrigerator with probability $\frac{1}{3}$, we find that the expected value of switching is

$$E_{\text{switch}}(X) = \frac{1}{3}(5{,}150) + \frac{1}{3}(7{,}650) + \frac{1}{3}(10{,}075) \approx 7{,}625.$$

This, sadly, is less than the expected value of sticking. In the long run we will do better by uniformly sticking on those occasions where Monty reveals a car.

The comparable calculation when Monty reveals a motorcycle goes as follows: If our initial door concealed a car, then our expected value works out to be 7,650. (Note that this scenario is equivalent to the second calculation above.) If our initial choice concealed a motorcycle, then our expected value is 10,150. And if our initial choice concealed a refrigerator, then our expected value is 12,575. Putting everything together now gives us

$$E_{\text{switch}} = \frac{1}{3}(7{,}650) + \frac{1}{3}(10{,}150) + \frac{1}{3}(12{,}575) = 10{,}125.$$

Since this is greater than the expected value of sticking (albeit not by very much), we see that it is wise to switch in this case.

The calculations when Monty reveals a refrigerator are very similar. In this case we have

$$E_{\text{switch}} = \frac{1}{3}(10{,}075) + \frac{1}{3}(12{,}575) + \frac{1}{3}(15{,}000) = 12{,}550.$$

It is in this scenario that we find the greatest advantage in switching.

Let us now consider the generalized form of this scenario given in version seven. We are given m types of prize with values $v_1, v_2, \ldots, v_m$. We further assume there are n_i prizes of value v_i for each value of i between 1 and m. Finally, by allowing prizes of zero value, we can assume there are as many prizes as doors.

We begin our analysis with a consideration of the sticking strategy. Our initial choice contains a prize of value v_i with probability $\frac{n_i}{m}$. It follows that

$$E_{\text{stick}} = \frac{n_1 v_1}{m} + \frac{n_2 v_2}{m} + \ldots + \frac{n_m v_m}{m} = \frac{1}{m}(n_1 v_1 + n_2 v_2 + \ldots + n_m v_m).$$

To simplify the notation, let us define

$$t = n_1 v_1 + n_2 v_2 + \ldots + n_m v_m.$$

Then the expected value of sticking is $\frac{t}{m}$.

As before, the expected value of switching depends on both the value of the prize revealed by Monty and the value of our initial door. Let us assume that Monty reveals a prize of value v_r and that our initial door conceals a prize of value v_i. Keep in mind that the remaining doors have an equal probability of concealing any of the remaining prizes. It follows that the expected value of switching is simply the average value of the prizes remaining in play different from both the prize Monty revealed and the prize behind your door.

Furthermore, following the procedure used above, we can write

$$E_{\text{switch}} = \sum_{i=1}^{m} P(\text{First choice has value } v_i)$$

$$\times E_{\text{switch}}(\text{Given that your first choice has value } v_i).$$

The probability that your initial choice has value v_i is $\frac{n_i}{m}$. The expected value of switching given that your initial choice has value v_i and Monty revealed a prize of value v_r is given by

$$\frac{n_1 v_1 + n_2 v_2 + \ldots + n_m v_m - v_i - v_r}{m - 2} = \frac{t - v_i - v_r}{m - 2}.$$

In going through the following computation, keep in mind that

$$\sum_{i=1}^{m} n_i = n_1 + n_2 + \ldots + n_m = m.$$

We now compute

$$E_{\text{switch}} = \sum_{i=1}^{m} \left(\frac{n_i}{m}\right)\left(\frac{t - v_i - v_r}{m - 2}\right)$$

$$= \frac{1}{m(m-2)}\left(mt - mv_r - \sum_{i=1}^{m} n_i v_i\right)$$

Since the expression in the parentheses is equal to $mt - mv_r - t$, the expression on the right simplifies to

$$E_{\text{switch}} = \frac{t}{m}\left(\frac{m-1}{m-2}\right) - \frac{v_r}{m-2}.$$

We find that this expression is equal to $\frac{t}{m}$ (the expected value of sticking), when $v_r = \frac{t}{m}$. This fraction represents the average value of all the prizes. Thus, if the value of the prize Monty reveals is equal to the average value of all the prizes, then it does not matter whether you stick or switch. If Monty's prize has a value smaller than the average, then you should switch. And if his prize has a value greater than the average, then you should stick.

We may as well finish this section with the obvious next question. Everything is as before, but this time we allow Monty to open an arbitrary number s of doors, with $1 \leq s \leq m - 2$. We still assume that Monty chooses his doors randomly, without regard for what is behind them.

The calculation will proceed in a manner nearly identical to the above. This time, suppose that Monty reveals a total of s prizes with a total value of x. We still have that the expected value of sticking is $\frac{t}{m}$. This time, however, after Monty does his thing we have $m - s - 1$ doors remaining in play that are different from our current choice. The total value of the remaining prizes is now $t - v_i - x$. It follows that our equation for E_{switch} becomes

$$E_{switch} = \sum_{i=1}^{m} \left(\frac{n_i}{m}\right) \left(\frac{t - v_i - x}{m - s - 1}\right)$$

$$= \frac{1}{m} \left(\frac{1}{m - s - 1}\right) \left(mt - mx - \sum_{i=1}^{m} n_i v_i\right)$$

$$= \frac{tm - t - xm}{m(m - s - 1)}.$$

The condition under which $E_{switch} = E_{stick}$ is now seen to be that $\frac{x}{s} = \frac{t}{m}$. That is, if the average value of the prizes Monty reveals is equal to the average values of the prizes generally, then it makes no difference whether you switch or stick. A pleasing generalization of our earlier finding.

5.7. Quantum Monty

Physicists have likewise taken notice of the Monty Hall problem, and the professional literature records several quantum mechanical versions of the basic game. Consider [16], [101], and the references contained therein, for example. Unfortunately, the level of mathematics and physics involved in presenting the details of these versions is far beyond anything we can discuss in this book. That inconvenience notwithstanding, a few words ought to be said about what the physicists are up to.

In section 4.7 we took a brief look at information theory. Our concern there was Shannon's insight that the amount of information received upon learning that an event has taken place is inversely related to probability; we learn more from hearing that a highly improbable event has taken place than we do from learning that something commonplace has occurred. This, however, is not the only issue with which information theorists concern themselves. There is also the problem of storing and manipulating information in efficient ways.

Everyday experience tells us that information may be stored in any physical medium that can exist in at least two distinct states. Computers, for example, store information in the form of sequences of zeros and ones. By altering the precise sequence of zeros and ones stored in the computer's memory, we also alter the information recorded there.

Storing information in this way requires assuming, first, that each digit in the sequence has a definite value, either zero or one. There is no bizarre intermediate state where the digit is simultaneously partially zero and partially one. Furthermore, any given place in the sequence records a zero or one independent of what we know about the sequence. I may not know whether, say, the fifteenth digit in the sequence is a zero or one, but it has a definite value nevertheless. If I examine the sequence and look specifically at the fifteenth digit, I will then learn a fact I did not previously know (i.e., whether that digit is zero or one). My examination will disclose to me the state of a certain physical system, but it will not by itself cause any change in that system.

We further assume that the value of any particular digit in the sequence is independent of all the other digits. Knowing that the fifteenth digit is a zero does not tell me the value of the forty-third digit, for example. Were this not the case, the quantity of information I could store would be severely compromised. If the value of the forty-third digit were determined by the value of the fifteenth, then I could not use the forty-third digit to store any new information.

These ideas feel so natural they hardly seem like assumptions at all. Just think about a well-shuffled deck of cards. I will not know whether the fifteenth card is red or black until I actually examine it. But there is no question that the card is either red or black independent of my state of knowledge. It is not as if the deck consists of completely blank cards, which then have suits and denominations appear from nowhere as soon as I take a look. Equally obvious is that the color of the fifteenth card in no way determines the color of the forty-third card. Of course these colors are independent. How could it be otherwise?

These assertions, obvious to the point of being trivial for large-scale physical systems, are simply false for systems the size of an atom. Physicists have devoted quite a lot of time to studying atoms, and the data they have collected can be explained only by abandoning the dictates of common sense. Consider, for example, the unstable nucleus of a radioactive atom. Suppose we place such a nucleus into a box and allow a minute to pass. We now ask, "Has the nucleus decayed in the past minute?" Until you open the box and check, you will not know how to respond. You would think, however, that the question has a definite answer even before you check. Either the nucleus has decayed or it has not, right? Yet this turns out to be incorrect. If the findings of quantum mechanics are to be believed, there is no definite answer to this question until we open the box and look. Prior to that examination, the nucleus exists in a so-called superposition of states, simultaneously decayed and not decayed. Such superpositions are ubiquitous among subatomic particles.

Then there is the phenomenon of entanglement. Under certain circumstances the attributes of one subatomic particle can be correlated with those of another in such a way that knowledge of one immediately gives you knowledge of the other. For example, entangled subatomic particles might be such that if we discover that one of them has an upward spin, we know immediately that the other has a downward spin, regardless of any physical separation between them. By itself this is not so puzzling. Consider, though, what we said in the previous paragraph. Like the simultaneously decayed and not-decayed nucleus, the subatomic particle does not have a definite spin until we actually take a measurement. Prior to taking the measurement there is only a superposition of all possible spins. Nonetheless, upon measuring the first we learn something about the second. But how does the second particle know the first has been measured?

There is nothing like a superposition of states or quantum entanglement in our everyday experience with relatively large objects. These are precisely the sorts of considerations that lead many physicists to speak of "quantum weirdness." And if our commonsense notions of the behavior of matter go out the window in considering subatomic particles, so too do our intuitions about information. It was from this realization that the field of quantum information theory was born.

Does all this strike you as strange? Counterintuitive? Preposterous, even? If it does, then you are not alone. Most physicists share your feelings. Quantum information theory is an entirely different beast from what we are used to, and even professionals in the field can find it difficult to keep their bearings. What is needed is a sort of anchor, something that will allow us to keep one foot in more familiar conceptual territory while we explore this new terrain.

And that is where the Monty Hall problem comes in.

Careful study of the classical problem revealed much regarding the nature of probability and classical information. At first the problem is beyond the intuition of all but the most savvy customers, but it rewards the effort put into understanding its subtle points. By finding quantum mechanical analogs for each aspect of the problem, we create an environment for studying the nature of quantum information in which we have at least some intuition to guide us. Quantum Monty Hall problems are not studied to help us determine optimal strategies in subatomic game shows. Rather, the insight gained from studying them aids us in applying modern physics to more practical pursuits.

The history of mathematics and science records many instances of this. Probability theory is nowadays an indispensable part of many branches of science, but for centuries it was studied solely in the context of gambling and games of chance. In the early days of computer science and artificial intelligence much emphasis was placed on the relatively unimportant problem of programming a computer to play chess. Insight into difficult mathematical or scientific problems often begins with the earnest consideration of trivial pursuits.

We have already seen how the Monty Hall problem opens the door to virtually every aspect of elementary probability theory. The next two chapters will describe the impact of the problem among cognitive scientists, psychologists, and philosophers. Now it seems it is helping shed some light on issues in quantum physics. Is there nothing the Monty Hall problem cannot do?

5.8. Monty at the Card Table

It is time for lighter fare. Writing in [57], bridge expert Phil Martin notes that Monty Hall–style errors in probabilistic reasoning sometimes occur at the card table. Even if bridge is not your thing, I urge you to stick around for the punch line at the end of this section.

We begin with a whirlwind tour of the rules of bridge. I will ask the card players in the audience to forgive me for the omissions in the following description. My intention is to explain only the really crucial points of the game, not to provide a primer on the subject. In particular, the notion of a "trump suit" will not be needed for our discussion, and we shall omit it.

Bridge is played by four players divided into two groups of two. One of the pairs is referred to as North-South, while the other is East-West. Each player is dealt a hand of thirteen cards. The play is now divided into two phases. The first is referred to as the auction, while the second is called the play. Throughout both phases each player tries to make reasonable inferences regarding the cards possessed by the other players.

Since the play is easier to explain than the auction, we will begin there. Playing in succession, each of the four players plays one card from his hand. All are required to play a card of the same suit as the first player. If a player does not have any cards of that suit, he may play any card of his choosing. The player discarding the highest card of the suit in play is said to "win the trick," and he starts the process over again by playing a new card. This phase of play is rather like the card game War that most of us played as children.

During the auction phase players bid for the right to control the initial play, and to establish a contract that they must fulfill under pain of suffering steep penalties for failure to do so. The contract stipulates the number of tricks the partnership must win. Relevant for us is the fact that a player uses his bids in part to signal to his partner the sorts of cards that he has. By the same token, each partnership listens to the bids of the other pair to draw information regarding the nature of their cards. The play during the early tricks provides further information concerning who has what cards.

Based on the manner in which the bidding plays out, the auction concludes with one of the players being made the "declarer." He is the one who begins the play and bears the burden of making his contract. His partner then becomes the unfortunately named "dummy." The dummy's cards are turned face-up on the table for everyone to see.

Now, place yourself in the role of the declarer, playing North. For strategic reasons you seek the location of, say, the queen of clubs. Since you know both your own cards and those of your partner (playing South), you can be certain that the queen is held by either East or West. You know that on the previous trick West began with a low spade, and his partner followed suit by also playing a spade. Based on the history of the game to this point and on certain probabilistic considerations, you have concluded that West holds five spades while East holds three.

You might now reason as follows: "If I am correct about the breakdown of spades, then East started with ten unknown cards (his original thirteen minus the three spades we think he has), while West started with eight (the five spades deducted from his original thirteen). Since any given card has an equal probability of being either in East's hand or West's hand, I conclude that the odds are five to four that East has the queen." In other words, based in part on the fact that West chose to lead a spade, we conclude that it is now more likely that East, rather than West, holds the queen.

Put another way, initially we think it is equally likely for any given card to be held by either of the two players. But now that we have seen West lead a spade, we think that any random club is more likely to be held by East rather than West.

Martin refers to this as "falling for the Monty Hall trap," for it does not fully consider all of the information we have. You see, if West had chosen to play spades at random and we knew somehow that this randomly chosen suit split five-three between West and East, then our reasoning would be valid. But West did not choose spades at random. It is typical, for strategic reasons, to lead a card in your longest suit (that is, the suit in which you have the most cards). Furthermore, probabilistically it is very likely that West will have at least one five-card suit. Since we knew ahead of time that West was likely to lead his longest suit, why should learning that that suit is spades alter the probability that West possesses some random club?

In the Monty Hall game many people mistakenly believe that Monty's actions cause the probability of our initial door choice to rise to $\frac{1}{2}$. The reality is that Monty's door-opening procedure cannot change the probability of our initial door. So it is here. Our prior knowledge of West's cards and the manner in which he is likely to choose his suit to play makes it unreasonable to think his decision to play spades alters the probability that he holds a random club.

I showed you this for two reasons. One is simply that it is a fascinating application of Monty Hall–style reasoning in an unexpected place. The other is this: Martin originally published this essay in *Bridge Today* magazine in 1989, before Marilyn vos Savant wrote her famous column. It was subsequently anthologized, in 1993, in a collection of high-level essays about bridge. The original essay began with a statement of the problem (previously presented in the Appendix to Chapter 1). Martin then wrote of the game show scenario, "Here the trap is easy to spot. But the same trap can crop up more subtly in a bridge setting."

The anthologized version contained the following footnote:

Easy to spot? Ha! About a year after this article was first published, the Monty Hall problem appeared in Marilyn vos Savant's column in *Parade* magazine. Ms. Savant stated, as I had, that the probability of your having chosen the grand prize remained one third. She received more than 1,000 letters scoffing at this conclusion and demanding that she print a "correction" and an apology. Many letters were from preeminent mathematicians and scientists. Eventually, the debate found its way to the front page of *The New York Times* (July 21, 1991). In the end, the scoffers had to eat their words. But I forever lost my right to call this problem easy.

Somehow that makes me smile every time I read it.

5.9. Literary Monty

One of my most pleasantly unexpected encounters with the Monty Hall problem came when I was reading the novel *The Curious Incident of the Dog in the Night-time*, written by Mark Haddon and published in 2003. The story is a first-person account of a teenager with Asperger's syndrome (a low-grade sort of autism) investigating the murder of his neighbor's dog. The narrator's condition leaves him with certain impairments in dealing with social situations. However, he has considerable facility with and interest in mathematics, and frequently pauses the action to discuss various mathematical tidbits. Indeed, a friend had recommended the book to me for that reason.

You can imagine my surprise, however, when I discovered an impressively lucid discussion of the Monty Hall problem splashed across pages 61–65. Haddon has his narrator describe the controversy that erupted as a result of Marilyn vos Savant's column. He then provides two different solutions to the problem. The first is the tree-diagram approach I explained in Chapter 2. The other involved conditional probabilities, and since I have not presented it previously I shall do so now. Note that the symbol $\wedge$ in what follows is being used to denote the word "and."

Let the doors be called X, Y and Z.
Let C_X be the event that the car is behind door X and so on.
Let H_X be the event that the host opens door X and so on.
Supposing that you choose door X, the possibility that you win a car if you then switch your choice is given by the following formula

$$P(H_Z \wedge C_Y) + P(H_Y \wedge C_Z) = P(C_Y)P(H_Z|C_Y) + P(C_Z)P(H_Y|C_Z)$$

$$= \left(\frac{1}{3}\right)(1) + \left(\frac{1}{3}\right)(1) = \frac{2}{3}.$$

I certainly never expected to find the Monty Hall problem spang in the middle of my pleasure reading, and it was especially nice to see such an elegant and novel approach. The problem is everywhere, folks. Ignore it at your peril.

6

Cognitive Monty

Though I have been a devotee of the Monty Hall problem for many years, I confess to being taken aback by the sheer volume of professional literature produced on the subject. Prior to beginning my research for this book, my file contained a handful of entries from various mathematics and statistics journals, but it had not yet occurred to me to search for material in other academic disciplines. So you can imagine my surprise upon discovering that philosophers, psychologists, economists, physicists, cognitive scientists, and many others besides had left their own mark on the problem. Tracking down all of this material was an enjoyable but highly time consuming task.

My collection of professional articles on the problem now contains over one hundred items, and I am quite certain I shall discover a dozen more as soon as this book is published. I found these articles by typing "Monty Hall" and sundry other strings into various databases and search engines. Most of these articles proved to be readily available via the Internet, and the rest were easily obtained through other means. The bibliographies of these papers led me to further papers published in even more obscure journals, many of which were nonetheless available online.

Of course, online resources do have their pitfalls. One article discovered via a search of the PubMed database (for articles in medicine and the life sciences) bore the uninspiring title "Eye Movement Responses of Heroin Addicts and Controls During Word and Object Recognition." This, remember, in a search on the string "Monty Hall." The abstract said much about heroin addiction and sensory capacity but little about game shows or elementary probability. The article was published in the journal *Neuropharmacology*, which only increased my puzzlement. A few mouse clicks later and the paper was on the screen in front of me. And that was when I noticed that the first two authors were R. A. Monty and R. J. Hall.

We begin our two-chapter tour of the academic literature on the Monty Hall problem with a consideration of what cognitive scientists and decision theorists have to say about it.

6.1. How Bad Is It?

Are things really so bad? Is the human brain, for all its magnificence, to be laid low by an elementary brainteaser derived from a game show? When we say that many people get the problem wrong upon hearing it for the first time, just how many people are we talking about?

In their 1995 paper "The Monty Hall Dilemma" [39], Douglas Granberg and Thad Brown put that question to the test. In a series of experiments they tested people's proclivities when confronted with the problem.

In one of their experiments subjects were assigned to one of three groups. Each group was told that they would engage in fifty trials of the basic Monty Hall game. They were also told they would accumulate points depending on how frequently they won the prize. The difference between the groups lay in the manner in which the points were awarded. The first group received one point for winning the prize at the end and zero points for not finding the prize. The second group likewise received no points for failing to find the prize. They received one point if they won the prize by sticking with their initial choice but two points if they won by switching. The intent was to provide an inducement to switching. This inducement was ramped up further in the final group, which received one point for a win by sticking and four points for a win by switching. As always, no points were awarded for failing to find the prize.

The results? In the first group 90% of the people stuck with their initial choice on their first play of the game, while 10% switched. Pretty grim. The second and third groups fared somewhat better, with 43% of the people in the second group and 36% of the people in the third group electing to switch on their first trial. (This difference was not found to be statistically significant.)

A tendency toward switching developed as the participants engaged in further trials of the game. Looking now at the final ten trials, we find 55% of the decisions made by members of the first group were to switch. The comparable numbers in groups two and three were 73% and 88% respectively. We further note that the percentage of people opting to switch every time in their final ten trials were 7, 17, and 39 respectively for the three groups. The implication is that people gradually learned through trial and error the benefits of switching but still opted to stick in many cases.

Granberg and Brown sum up their findings as follows:

Why do people have such a difficult time learning inductively to switch across 50 trials? To us, having worked with the computer program used in these experiments, switching in the MHD is now an obvious, almost trivial strategy. One of us even worried that the experiment would be too apparent and that midway through the sequence of trials, subjects would surely see the solution and begin to switch with little variance. Yet even for those with the greater incentives to play a switching strategy, full insight into the correct solution and purely rational behavior did not emerge.

Numerous other researchers have carried out similar experiments. Surveying these studies in 2004 [12]. Burns and Wieth report:

These previous articles reported 13 studies using standard versions of the MHD, and switch rates ranged from 9% to 23% with a mean of 14.5% ($SD = 4.5$). This consistency is remarkable given that these studies range across large differences in the wording of the problem, different methods of presentation, and different languages and cultures. Thus it appears that failure on the MHD is a robust phenomenon unlikely to be due to confusion arising from minor aspects of the wording or presentation of the problem.

Pretty bleak. Just writing this section has gotten me down. Let us move on.

6.2. The Virtues of Sticking

There is something puzzling about the data reported in the last section. We know that a proper mathematical analysis of the Monty Hall problems leads to the conclusion that you win two-thirds of the time by switching. The most popular incorrect argument asserts that after Monty eliminates a door, the remaining two are equally likely. This leads to the conclusion that there is no advantage to be gained by switching doors.

Why, then, do most people prefer sticking?

In the experiment described in the previous section, fully 90% of the people in the group that was not given an incentive to switch elected to stick on their first play of the game. You might think that, perceiving the remaining doors to be equally likely, some people would switch just for fun.

Back in Chapter 2 I mentioned that many of my students argue for sticking on the grounds that if you switch and then lose, you will feel so much worse than if you lose by sticking that it is best not to take the chance. There is a substantial literature in the cognitive science, psychology, and economics journals suggesting that this is a real tendency in human thought.

Specifically, a negative consequence incurred by inaction causes less regret than the same negative consequence incurred by action. For example, suppose Mr. Jones is considering selling one share of stock A for $10 and then using that money to purchase one share of stock B. If he opts not to make the sale and the next day company A goes out of business and his stock becomes worthless, he will feel very bad. But he will feel even worse if he makes the sale and then company B goes out of business, at least if the research in this area is to be believed. This despite incurring the same $10 loss in both cases.

Is that sort of reasoning implicated in the popularity of sticking in the Monty Hall problem? In [34], Gilovich, Medvec, and Chen investigated that question. In the game as played in this experiment, all of the subjects believed they had an opportunity to win either a grand prize or a modest prize. In reality the game was rigged so that no one received the grand prize. Some of the subjects were placed in groups where they would "lose" by sticking, while others would "lose" by switching. I will not belabor here the full details of the clever experimental procedure that created this situation. Feel free to consult the actual paper if you are interested, but be warned that the explanation occupies several pages.

Subjects were then asked to place a monetary value on the modest prize that they won. The hypothesis was that those who received the modest prize by switching would feel worse than those who received it by sticking. The switchers would consequently attempt to reduce their feelings of dissonance by valuing the modest prize more highly than the stickers. ("I didn't mess up *that* badly," they could argue. "This modest prize is pretty nice!")

The results were in accord with this prediction. I turn the floor over to the authors:

> It appears that people do indeed reduce dissonance more for their errors of commission than their errors of omission. Subjects who switched *from* a box that was subsequently found to contain a grand prize tended to value the modest prize they received more highly than those who failed to switch *to* the box containing the grand prize. . . .
>
> These results both complement and extend previous findings in the counterfactual thinking literature. Previous investigators have presented data indicating that people may feel more pain over a bad outcome that stems from action taken than an action forgone. What we have shown is that this initial sting of regrettable action can be undone by the process of dissonance reduction. Because action tends to depart from the norm more than inaction, the individual is likely to feel more personally responsible for an unfortunate action. Thus, subjects who switched boxes in our experiment were more likely to experience a sense of "I brought this on myself," or "This need not have happened," than subjects who decided to keep their initial box.

An interesting finding. It would seem that for many people the decision to stick in the Monty Hall problem results not just from a misapprehension of the probabilities but also from a variety of emotional considerations that are ultimately "irrational." Since there is no obvious advantage to switching, the desire of people to avoid taking action that leads to a negative consequence (as opposed to allowing the same negative consequence to occur via inaction) takes over and influences their decision.

This suggests the following experiment. For most people the advantage in switching becomes obvious as the number of doors goes up. That is, if the game is played with one hundred doors, with Monty eliminating ninety-eight goat-concealing doors after the player makes his initial choice, then virtually everyone recognizes that switching is the right decision. With just three doors, this is manifestly not the case. The question is this: At what point does the advantage in switching become sufficiently clear that it outweighs people's natural tendency to stick? If it is not obvious with three doors, then how about with four? Or five? Following the general rule that nothing is as it seems in the Monty Hall problem, I would not even hazard a guess as to the critical number.

A step in this direction was taken by economist Scott Page in [72]. He tested subjects in three different versions of the Monty Hall problem. The first was the classical version with three doors. In the second, subjects were confronted with ten doors. The third featured one hundred doors. Regardless of the number of doors at the start, Monty always opened all but one of the remaining doors after the subject made his initial choice, never opening a door that concealed the prize. Consistent with the results of other researchers, Page found that fewer than 12% of subjects opted to switch in the three-door case. That jumped to nearly 50% in the ten-door case and 95% in the one-hundred-door case. In a second experiment Page had subjects play the three-door and hundred-door versions of the game one right after the other. He found that even though most people understood to switch in the hundred-door case, they did not then transfer this intuition to the three-door case.

This confirms a point I made back in section 2.2.1. That it is obvious that one should switch in the hundred-door case does little to help resolve the difficulty of the three-door case.

6.3. Why All the Confusion?

We now move to [4], an entry from 1982 bearing the seductive title "Some Teasers Concerning Conditional Probability" and coauthored by Maya Bar-Hillel and Ruma Falk. Why, they wondered, do people find problems of conditional probability to be so confusing and paradoxical? Not at the level of formal mathematics, mind you. No one expects people with no particular training in mathematics to understand the technical details of Bayes' theorem. But why do people's intuitions so often lead them astray on problems of this sort, even after their errors are made plain via rigorous, logical arguments? Part of Bar-Hillel and Falk's discussion was given over to the problem of the three prisoners (hereafter referred to as the "prisoners' problem"). Since we are already well versed in the niceties of *that* little puzzle, we will ponder instead one of their other examples. It may be stated thus:

> Mr. Smith is the father of two. We meet him walking along the street with a young boy whom he proudly introduces as his son. What is the probability that Mr. Smith's other child is also a boy?

(Recall that we briefly mentioned this problem in section 1.10.)

Two possible arguments suggest themselves. In the first we note that the sex of one child can have no bearing on the sex of the other. Consequently, Mr. Smith's other child is either male or female, and each of these possibilities ought to be assigned a probability of $\frac{1}{2}$. In effect, we are arguing that the knowledge that Mr. Smith has one boy tells us merely that his children are either BB (for two boys) or BG (for one boy and one girl). The information available provides no reason for preferring one of these possibilities over the other, and so each gets a probability of $\frac{1}{2}$.

The second argument begins with the observation that prior to meeting Mr. Smith's son we have four possibilities. Taking birth order into consideration, we have BB, BG, GB, and GG. Upon meeting his son, we can eliminate only GG. The remaining three possibilities are equiprobable, and we therefore arrive at an answer of $\frac{1}{3}$. At this point the familiar Monty Hall vertigo sets in, as we are faced with two reasonable arguments that nonetheless point in opposite directions.

Part of the difficulty is a genuine vagueness in the problem. To see this, let us consider three variations on the basic scenario. In the first, we imagine that Mr. Smith tells us that the boy with whom he is walking is actually his eldest child. In the second, we learn that Mr. Smith chose his walking companion at random from among his two children. And in the third, we do not meet one of Mr. Smith's children at all. Instead we run into him walking down the street alone, and he says. "I have two children and at least one of them is a boy."

We consider the variations in turn. If Mr. Smith tells us specifically that his walking companion is his oldest son, then the problem effectively becomes the following: "Mr. Smith's oldest child is a boy. What is the probability that his youngest child is also a boy?" Put this way, it is clear we are in the scenario envisioned by the first argument above. Mr. Smith's children, listed from oldest to youngest, are either BB or BG. Since the sex of the first child has no bearing on the sex of the second, we arrive at a probability of $\frac{1}{2}$ in this case.

We next consider the possibility that Mr. Smith chose his walking companion randomly from among his two children. We could then reason, in accord with the proportionality principle discussed in Chapter 3, that if

Mr. Smith has two boys, it is a certainty that his walking companion will be male. On the other hand, if he has one boy and one girl, the probability of choosing a male walking companion is only $\frac{1}{2}$. It follows that the updated probability that Mr. Smith's children are BB must be twice that of either BG or GB. Since these three probabilities must add up to one, we obtain an updated probability of $\frac{1}{2}$ for BB (and of $\frac{1}{4}$ each for BG and GB). It would seem these first two variations have the same answer.

Move now to the third variation. We can reason as follows: among all the families with two children and at least one son, roughly one-third of them will be BB, while the other two-thirds will be either BG or GB. Thus, it is in this scenario, alone among the variations we have considered, that we are justified in assigning equal posterior probabilities to BB, BG, and GB, and since the second child is a boy in exactly one of them, we arrive at an answer of $\frac{1}{3}$.

An amusing problem, and one that teaches us an important lesson. In conditional probability problems, it is not just the new data that are important. Equally necessary is an understanding of the precise statistical experiment that led you to the data. All of the problems just discussed have the same basic form: You begin with the knowledge that Mr. Smith has two children. You then learn that one of them is a boy. You must then determine the probability that Mr. Smith has two boys. However, unless we know the precise manner in which we learned that Mr. Smith has a boy, we can justify several different answers.

From this analysis, and of similar analyses applied to other problems in conditional probability, the authors draw two lessons. Here is the first:

> The kind of problem in which the conditioning event does turn out to be identical to what is perceived as 'the information obtained' can only be found in textbooks. Consider a problem which asks for 'the probability of A given B.' This nonepistemic phrasing sidesteps the question of how the event B came to be known, since the term 'given' supplies the conditioning event, by definition. For example, the answer to the question: 'What is the probability that "Smith has two sons" *given* "Smith has at least one son"?' is (under the standard assumptions) unequivocally $\frac{1}{3}$. Outside the never-never land of textbooks, however, conditioning events are not handed out on silver platters. They have to be inferred, determined, extracted."

The second point relates to the reason people find the results of these problems so paradoxical. Consider once more the question of whether knowledge of the birth order of Mr. Smith's children is relevant to discerning the posterior probability of BB. On the one hand, such knowledge narrows the sample space from $\{BB, BG, GB\}$ to $\{BB, BG\}$. Viewed that way, it would seem to be highly relevant indeed. On the other hand, there is also a plausible argument for regarding it as irrelevant. Our determination of the probability of BB is the same regardless of whether Mr. Smith tells us it is his eldest child or his youngest child that is a boy. And if the probability of BB conditioned on two complementary events, such as "eldest" and "youngest," is the same, then the unconditional probability of BB must take on this same value.

Why are these results so confusing for most people? Because they place two strong intuitions at odds with each other. As exercises in formal mathematics, these problems yield to familiar techniques once the proper assumptions are spelled out in sufficient detail. The situation is quite different, however, when we encounter similar situations in real life. Our usual intuitions about updating past beliefs in the face of new evidence are not always adequate.

A nice paper, and one that presents an impressively lucid explanation of the manner in which the solutions to their familiar probability exercises differ depending on the assumptions that go into them. Their explanation for the dissonance set up between our intuitions and the rigorous solutions of these problems is certainly plausible as well. That said, I do fault Bar-Hillel and Falk for two things.

The first is their implication that authors of probability textbooks have overlooked something in presenting things the way that they do. Typically, however, such books present a rigorous, mathematical treatment of the material, and such a treatment inevitably takes place in a "never-never land," far removed from the vagaries of daily existence. That, after all, is why we speak of "abstract" or "pure" mathematics. The exercises presented in such books are intended not as primers on proper decision making in realistic situations, but merely as illustrative and simplified examples of how to make proper use of the often complex mathematical formalism presented in the text.

A second and somewhat more serious objection is the manner in which Bar-Hillel and Falk opt to make their central point. For example, they write, "We showed that different ways of obtaining the selfsame information can significantly alter the revision of probability contingent upon it." In the context of their article I see the point they are making. They are thinking of "the information" as "Mr. Smith has a boy," with the "different ways of obtaining" that information corresponding to the scenarios described above.

I suspect most mathematicians and statisticians would not see things that way. I, for one, would say that the problem as initially given, in which we simply see Mr. Smith walking with one of his children, is vague. It cannot be solved without bringing in further assumptions that are not explicitly justified by the statement of the problem. In the more detailed scenarios, however, we are not learning the same information ("Mr. Smith has a boy") each time. Rather, we have three different problems with three different pieces of information provided. Specifically, "Mr. Smith's eldest son is a boy," "A child chosen at random from Mr. Smith's two children is a boy," and "Mr. Smith told us that one of his children is a boy," respectively. For the purposes of mathematical modeling, the nature of the statistical experiment performed ought to be regarded as part of the information necessary for establishing the correct posterior probabilities.

6.4. Subjective Theorems

Starting in the late eighties and continuing to the present, the professional literature has seen a steady trickle of research dedicated to figuring out just why, exactly, people find the Monty Hall problem so confusing. An early representative of the genre is the paper by Shimojo and Ijikawa [83]. This having appeared prior to the fracas with Marilyn vos Savant, its focus is the prisoners' problem.

The core of their experiment rested with two different versions of the problem. The first was the classical version. Recall that this involves three prisoners, A, B, and C, awaiting execution. They know that one person, *chosen at random*, was to be set free, while the other two are to be executed. Prisoner A asks the jailer to tell him the name of one man who will be executed. The jailer replies that B will executed. We now ask whether this revelation alters A's chances of being set free. As we know, A's chances of being set free remain at $\frac{1}{3}$, while C's chances have now improved to $\frac{2}{3}$ (given, as always, certain assumptions about how the jailer decides which name to reveal). We will refer to this version as problem one.

The implication of the italicized phrase is that each prisoner begins with a $\frac{1}{3}$ chance of being executed. Altering that assumption fundamentally changes the nature of the problem. Shimojo and Ijikawa devised a second version in which the prisoners do not have an equal probability of being freed. Instead, the selection process was biased so that A and B had a probability of $\frac{1}{4}$ of being freed, while C had a probability of $\frac{1}{2}$. We are to assume that A is aware of this fact. We ask again how he ought to assess his chances of being freed upon being told that B is to be executed. This shall be referred to as problem two.

By now we are well versed in the use of Bayes' theorem in solving such problems. We will abuse notation by using A, B, and C to denote the events in which prisoners A, B, and C respectively are set free. We shall use a, b, and c to denote the events in which the jailer says that A, B, and C respectively are to be executed. We make the standard assumption that when the jailer can choose from among two different prisoners in answering A's question, his choice is made randomly (and, of course, that the jailer always tells the truth).

We then have the prior probabilities

$$P(A) = \frac{1}{4}, \quad P(B) = \frac{1}{4}, \quad P(C) = \frac{1}{2}.$$

We also have the conditional probabilities

$$P(b|A) = \frac{1}{2}, \quad P(b|B) = 0, \quad P(b|C) = 1.$$

We now employ Bayes' theorem to determine $P(A|b)$ as follows:

$$P(A|b) = \frac{P(A)P(b|A)}{P(b|A)P(A) + P(b|B)P(B) + P(b|C)P(C)}$$

$$= \frac{\left(\frac{1}{4}\right)\left(\frac{1}{2}\right)}{\left(\frac{1}{2}\right)\left(\frac{1}{4}\right) + 0\left(\frac{1}{4}\right) + 1\left(\frac{1}{2}\right)} = \frac{1}{5}.$$

It would seem, then, that A's chances of receiving the pardon have dropped. As noted by Shimojo and Ijikawa, this adds another layer of confusion to the problem, as many have a strong intuitive feeling that A's chance of being pardoned should not go down upon hearing the jailer's statement. The result is correct nonetheless.

Based on preliminary interviews with graduate students having some familiarity with basic statistics, the authors formulated an initial hypothesis regarding the reasoning used by people in solving such problems. The hypothesis involved the use of three "subjective theorems," by which the authors meant guiding principles that seemed reasonable and intuitive to the people employing them, but were not truly theorems in the mathematical sense. These subjective theorems, as they described them, were:

- *"Number of cases" theorem:* When the number of possible alternatives is N, the probability of each alternative is $\frac{1}{N}$.

- *"Constant ratio" theorem:* When one alternative is eliminated, the ratio of probabilities for the remaining alternatives is the same as the ratio of prior probabilities for them.

- *"Irrelevant, therefore invariant" theorem:* If it is certain that at least one of the several alternatives $(A_1, A_2, \ldots, A_k)$ will be eliminated, and the information specifying which alternative to be eliminated is given, it does not change the probability of the other alternatives $(A_{k+1}, \ldots, A_N)$.

Mind you, the claim was not that people would necessarily formulate their thinking in such precise terms. Instead it is simply that these theorems capture the main approaches used by people in pondering the problems.

Now for the really clever part. In problem two each of the four proposed solution methods (Bayes' theorem and the three subjective theorems) lead to different answers. The details are recorded in Table 6.1, where the fractions represent the value of $P(A|b)$ obtained for the two problems via each of the four methods.

Table 6.1: Solutions given by different approaches in variations of the prisoners' problem

Theorem	Problem One	Problem Two
Bayes' theorem	1/3	1/5
"Number of cases"	1/2	1/2
"Constant ratio"	1/2	1/3
"Irrelevant, therefore invariant"	1/3	1/4

To understand that second column, note that we have already worked out the correct answer of $\frac{1}{5}$ from Bayes' theorem. The "Number of cases" theorem would argue that after B has been eliminated, only two possibilities remain, and they ought to be regarded as equally likely. The "constant ratio" theorem argues that since prior to the jailer's revelation the ratio $P(A)/P(C)$ was equal to $\frac{1}{2}$, that ratio should be maintained after receiving the jailer's news. Since $P(A|b)$ and $P(C|b)$ must sum to 1, we obtain $P(A|b) = \frac{1}{3}$. The "irrelevant, therefore invariant" theorem argues that since we already knew that at least one of B or C would be executed, we have learned nothing important from the jailer's revelation. It follows that $P(A)$ should remain unchanged at $\frac{1}{4}$.

By a series of questionnaires featuring these and related problems, the authors were able to get a feel for how people approached things. In their conclusions they note first that both of the problems featured here proved to be intuitively difficult. Fully half the participants gave the wrong answer to problem one, and of those giving the correct answer nearly all offered incorrect reasoning. On problem two, almost nobody could determine whether A's chance of survival would increase or decrease after the jailer's revelation.

They further concluded that participants used a variety of subjective theorems in formulating their answers. Moreover, people tended to use the same subjective theorems regardless of the problem they were considering. The "number of cases" and "irrelevant, therefore invariant" theorems proved especially popular, showing up in the reasoning of more than 95% of the participants. Finally, people's approaches to the problems did not change significantly even when the problems were phrased in a way to suggest one method of approach over another.

The experiments of Shimojo and Ijikawa put meat on the bones of the suggestion by Bar-Hillel and Falk that people have conflicting intuitions when approaching these questions. Lacking the mathematical training to properly assess all of the relevant factors for updating the prior probability, people fall back on certain rules of thumb that are intuitively satisfying and often useful (just not for the problem at hand). The conflict between these intuitive rules and the proper Bayesian approach explains why the problems seem paradoxical to so many people.

6.5. Primary and Secondary Intuitions

In the 1992 paper [22], psychologist Ruma Falk picked up where Shimojo and Ijikawa left off. While agreeing that the subjective theorems presented in the last section are important, she refined things further by distinguishing between primary and secondary intuitions.

Among the primary intuitions are the "uniformity belief" and the "no news, no change belief." The uniformity belief is identical to the "number of cases" subjective theorem from Shimojo and Ijikawa. It holds that when one of a collection of equally probable alternatives is shown to be impossible, then the probability redistributes itself equally over the remaining possibilities. "No news, no change" is a small variation on the "irrelevant, therefore invariant" theorem. It holds simply that probabilities can only change in the face of new information. In the context of the prisoners' problem, since A knows ahead of time that one of B and C will be executed, he is not receiving any news from the jailer's disclosure. Therefore, his probability ought not to change. This, recall, is the argument used by A to persuade the jailer to reveal a name.

Falk observes, following Shimojo and Ijikawa, that these beliefs are ubiquitous in human reasoning about probability. In the case of the prisoners' problem, they point in different directions. She goes on to note that people avoid the conflict by clinging to one or the other of these beliefs even in the face of contrary evidence, with the "uniformity belief" seeming to win out by majority vote. Falk writes.

> Those believing in posterior uniformity are not only numerous, there is also a special quality to their conviction: they are highly confident. Their belief can be considered a *primary intuition* since it is marked by some of the most distinctive features of such a cognition. ... Indeed, people rarely display any shred of doubt when they instantaneously rely on the uniformity assumption, as if there is *intrinsic certainty* to that belief. Intuitive beliefs ... exert a *coercive* effect on the individual's reasoning and choice of strategy. Intuition is also characterized by *perseverance* in being resistant to alternative arguments.

As for the "no news" argument, Falk observes that even knowledgeable people sometimes fall back on it. In the present case it is made all the more persuasive by the fact that it gives the correct answer. Marilyn vos Savant used a variation of it when trying to persuade her angry correspondents of the value of switching. In explaining her reasoning in the Monty Hall problem, vos Savant described a game in which a pea is placed under one of three shells. The player then places his finger on one of the shells. Vos Savant writes:

> The odds that your choice contains a pea are $\frac{1}{3}$, agreed? Then I simply lift up an empty shell from the remaining two. As I can (and will) do

this regardless of what you've chosen, we've learned nothing to allow us to revise the odds on the shell under your finger.

The fallacies in these intuitive views are easy to spell out. The uniformity assumption overlooks the importance of assessing the conditional probabilities of receiving the new information given each of the other possibilities in turn. And the no-news argument is fine as far as it goes, except that in practical situations it is often difficult to discern, absent a thorough Bayesian analysis and a meticulous attention to detail, whether or not we have actually learned anything.

By secondary beliefs Falk envisions "semi-intuitive heuristics that are arrived at through some deliberations, and which seem plausible once they are formulated. Secondary intuitive beliefs are acquired partly as a result of instructional intervention." She offers three examples of such beliefs: (1) the constant-ratio belief, (2) the symmetry heuristic, and (3) the likelihood-ratio heuristic.

Falk provides a thorough discussion of each of these, but we will be content simply to give brief definitions. Shimojo and Ijikawa included the constant-ratio belief among their subjective theorems, so we will say no more about it here. In the context of the prisoners' problem, the symmetry belief holds that prior to the jailer's statement, A should regard his fellow prisoners, B and C, as being symmetric. That is, he will reason in the same way regardless of which specific person the jailer names. Consequently, actually hearing a specific name gives him no reason to alter his assessment of his chances. Finally, the likelihood principle supposes that you are given complementary events A and $\overline{A}$ (in this case that A will or will not go free). You then learn piece of information b (that the jailer has named prisoner B as one who will definitely be executed). Then $P(A|b)$ is greater than $P(A)$ precisely when $P(A|b)$ is greater than $P(\overline{A}|b)$. This can be seen as a misapplication of Bayes' theorem.

Just to be clear, the point is not that people who aren't mathematically trained formulate their thinking in these terms. Rather, it is that these ideas seem to arise frequently in discussions of these problems. All of these secondary intuitions are plausible, and in many situations they lead to accurate conclusions. As general rules, however, they fall flat.

The picture that emerges from the papers of Bar-Hillel and Falk, Shimojo and Ijikawa, and Falk is that of people trying to solve subtle problems using faulty, all-purpose shortcuts. They are applying a sledgehammer to a problem that requires a scalpel. It seems there is something in our cognitive architecture that leads us to make fools of ourselves when discussing problems of this sort.

Two sections from now we will consider some challenges to this general view. First, however, after all our hard work in the previous sections, let us ponder something a bit less dense.

6.6. Monty Goes International

If you are thinking that incomprehension of the Monty Hall problem is strictly an American phenomenon, then I urge you to think again. Sociologist Donald Granberg investigated that possibility. His findings were reported in [36] in 1999.

Specifically, Granberg investigated the following question:

> To the U.S. subjects tested in the initial MHD [Monty Hall dilemma] studies, this solution seemed highly counterintuitive; consequently they showed a strong tendency to stick when they should have switched. This raises a question that can be addressed only through cross-cultural comparisons. Is there something inherent in the intersection between human cognition and the MHD itself which leads people generally to stick in this two-stage decision when they should switch? Or is there something specific to the socialization process in the U.S. which leads people reared in the U.S. to respond predominantly in an incorrect way to the MHD?

Granberg and his associates then presented the Monty Hall problem to college students from the United States, Sweden, China, and Brazil. The article explains that these countries were chosen on the basis of "the author's available contacts." The students were also presented with a variation on the basic Monty Hall scenario, but I will not discuss that here.

The results?

> In each country people responded similarly to the MHD by showing a strong tendency toward the nonoptimal answer of sticking with their initial selection. The percentage of subjects who checked "stick" in the MHD overall was 83%, and ranged between only 79% for China and 87% for Brazil. The U.S. and Sweden were intermediate, with 84% and 83% respectively, checking that they would stick.

While further research in this area might be nice (for example, do people from different cultures nonetheless make use of the same subjective theorems noted by Shimojo and Ijikawa?), Granberg's data suggest that confusion over the Monty Hall problem is telling us something about human psychology. I will give the last word to Granberg:

> Our main finding is cross-cultural similarity in the tendency of people to respond erroneously to the MHD type of two-stage decision problem. Thus, it appears that there is something inherent in the MHD, as a cognitive illusion, which leads people in very different cultures to respond similarly. People misapprehend the true probabilities,

perceiving the odds to be even, and then tend to stick with their original selection. ... Because the four cultures studied encompass very considerable variety, the temptation is to infer that the tendency to stick in the MHD reflects a universal human propensity; it may, however, be premature to go that far.

6.7. Humanity Fights Back

Are things really that bad? Is the human brain really so inadequate that a straightforward game-show brainteaser sends it into a funk? Surely there is more to the story than *that*.

There have been at least two attempts to mitigate the damage to human self-esteem dealt by the research previously described. In one it is stressed that the solutions to probabilistic problems of the sort under consideration in this chapter depend critically on often unstated assumptions. That being the case, the implications of research in which people give only their answers to probabilistic problems, and not their reasoning, are now unclear. One possibility is that people are precisely the poor probabilistic reasoners they appear to be. But another is that their reasoning is perfectly sound; it only appears otherwise because they are making assumptions different from what the experimenter envisions.

Representative of this genre is a contribution by Raymond Nickerson [67]. The first half of this lengthy paper is given over to an unusually clear and thorough presentation of a number of probabilistic brainteasers, including the sibling gender problem and the Monty Hall problem. Nickerson does a fine job of explaining the details of how different assumptions lead to different answers in these problems. In fact, in a few places he might have been a bit too thorough. He discusses a version of the sibling gender problem first offered by Martin Gardner: "Mr. Smith says 'I have two children and at least one of them is a boy.' What is the probability that the other child is a boy?" Gardner answered that the probability is $\frac{1}{3}$, on the grounds that the three possibilities BB, BG, and GB are now equally likely. Bar-Hillel and Falk [4] also discussed this example and endorsed Gardner's reasoning, as did I, you will recall, a few sections ago. Nickerson writes,

> There is, however, an unstated assumption in this reasoning, namely that Smith will, with $p = 1$, report that he has at least one son if he has at least one son. An alternative possibility is that if he has one son and one daughter, he will, with some probability less than 1, report that he has at least one son and, with the complementary probability, report that he has at least one daughter. On this assumption, the answer to the question would not be $\frac{1}{3}$ but could be anything between $\frac{1}{3}$ and 1. It would be $\frac{1}{2}$ if Smith were equally as likely to report having at least one

son as to report having one daughter when he had both, and it would be 1 if Smith reported having at least one daughter if he could possibly do so.

Yes, well, if *that's* how you want to play it, we also have to assume that Mr. Smith does not have any sexually ambiguous children. And that his perceptions of his children's sexes are always accurate. There really does come a time when the hair-splitting must stop, unless we specifically *want* to drive people insane with these problems.

Kidding aside, I can think of a number of journals in mathematics and statistics that would have been happy to publish the first half of Nickerson's article. It really is an excellent survey of the mathematical issues raised by these problems—better, frankly, than what most mathematicians would have written. I must, however, object to one of his conclusions:

> It is often the case that the descriptions of situations used in studies of statistical or probabilistic reasoning leave important details unspecified; when this is the case, participants have to make assumptions, at least tacitly, to proceed. Without knowing what assumptions they have made, conclusions about the rationality or irrationality of their behavior are on shaky ground.

This, I fear, is wishful thinking, at least with regard to the Monty Hall problem. I base this conclusion on two pieces of evidence, one formal and one informal. The informal evidence is my own experience in presenting this problem to my students. With calculus students it simply never happens that anyone asks me to clarify some subtle point in the problem's statement. (Yes, I present the Monty Hall problem to calculus students. There are some days when you just can't stomach the idea of working out still more examples of integration by parts.) And when I call their attention to the finer points of the problem, I typically get puzzled looks for my trouble. The grim statistics improve only slightly when presenting the problem to math majors. (Even here, one suspects the improvement is not so much the result of improved probabilistic reasoning but rather the result of math majors being savvy enough to realize that I would not have asked the question if the obvious answer were correct.)

The formal piece of evidence lies with the studies that have been done on precisely this point. An especially nice example is the 1990 paper by Ichikawa and Takeichi [46], unreferenced by Nickerson. The point about problem ambiguities affecting the responses people give to these problems had been raised previously by certain Japanese researchers. Ichikawa and Takeichi describe the situation this way:

> Shimojo and Ichikawa's (1989) explanation to the difficulty of the problem suggests human erroneous beliefs on the nature of probability.

However, if Ihara's or Ito's account is to the point, the difficulty is an artifact and human belief system is congruent with the probability theory. From our viewpoint, the difficulty of the problem does not seem to be accounted for by the ambiguity of the jailer's choice probability.

They then devised an experiment specifically to determine the role played by various assumptions in people's reasoning when confronted with these problems. It would take us too far afield to discuss the minutiae of the experimental design. Suffice it to say that their results plainly supported the idea that it is a failure of human probabilistic reasoning that these problems seem so difficult, and not the result of ambiguities in the problem statement.

People are not getting hung up over the issue of subtle, unstated assumptions. The issue is people applying intuitive arguments to unfamiliar problems that can only be solved properly by more complex means. It is telling us something significant about human cognition that people have such difficulty reasoning properly about uncertainty. We should not hide from that conclusion.

The second line of defense has been ably expressed by Aaron and Spivey-Knowlton in [1]. The difficulty with these problems, they suggest, lies not in the fundamental inability of human beings to come to terms with probabilistic reasoning. Rather, the issue, at least in part, is the manner in which these problems are expressed. They write:

> Recently, however, Gigerenzer has suggested an explanation for human performance on these tasks without claiming that people lack the ability to function as Bayesian agents. People do have methods or *algorithms* for reasoning about probabilities, but as humans evolved over the ages, the algorithms for Bayesian reasoning were not exposed to information expressed as probabilities. Instead, people gathered information as it came to them, one event at a time and not with the collective information about a set of events that probabilities would give. Thus Gigerenzer suggests in his framework of *Ecological Intelligence* that people's Bayesian algorithms are adapted for *natural frequencies* (e.g. "out of 160 coin tosses, 80 landed heads") as opposed to the probabilities (e.g. "50% of the coin tosses landed heads") in which information is traditionally presented in studies that produce anti-normative evidence. This difference between *information formats* may not seem dramatic, but in some contexts it can have important effects.

The implication is that at least some of the difficulty people have with the Monty Hall problem could be ameliorated by presenting the problem with frequentist, as opposed to probabilistic, language.

To test this idea, Aaron and Spivey-Knowlton presented participants with two different questionnaires related to the Monty Hall problem. They began with identical statements of the problem. The first questionnaire then asked participants to imagine that 30,000 rounds of the game have been played in which the player chooses door *A* every time. They were then asked questions such as, "Of these 30,000 rounds in which the player chooses door *A* in part 1 of the round, in how many is the car actually behind door *A*? Of the rounds in the answer to question 1, the rounds in which the player chooses *A* in part one and the car is actually behind door *A*, in how many of those rounds will the host open door *B* in part 2 of the round?" This continued through a series of eleven questions, the idea being to force participants not just to work out whether it is wise to switch or stick, but also to carry out various frequency calculations.

The second questionnaire eschewed frequentist language, preferring its probabilistic counterpart. Thus, the analog of the questions from the previous paragraph were. "In a round in which the player chooses door *A* in part 1, what is the probability that the car is actually behind door *A*? In a round as in question 1, in which the player chooses door *A* in part 1 and the car is actually behind door *A*, what is the probability that the host will open door *B* in part 2 of the round?"

The results?

> The results compellingly demonstrate that information presented and manipulated in a frequency format facilitated Bayesian competence in understanding the mathematics underlying the three doors problem. Given a presentation of the problem in frequency format, rates of correctness on the math questions ranged from 7% to 26%, depending on the experiment. Given a presentation in probability format, correctness on the math questions was a flat 0% in all experiments. This supports Gigerenzer's hypothesis about the importance of information format to normative Bayesian performance in inference tasks.

An interesting finding, and one that I easily believe. Though I have been seriously thinking about this problem for quite some time, I found the frequentist questions far easier to answer than the probabilistic ones. Perhaps people do find it easier to deal with concrete numbers than abstract computations. This finding certainly suggests that in trying to persuade people of the advantage in switching, a frequentist approach might be the way to go.

That said, I do not see in this experiment anything to affect our view of the previous experiments on which I have reported. That people can be made to understand what is going on by a sufficiently careful and concrete presentation does not mitigate the fact that in its classical presentation the Monty Hall problem is baffling to nearly everyone. Furthermore, even in the frequentist version the percentage of people answering correctly was depressingly low. Apparently even when the problem is presented in a way specifically designed to appeal to people's natural intuitions, most still find it confusing.

The experiment by Aaron and Spivey-Knowlton could easily underwrite the conclusion that many people are even worse at probability than previously imagined.

6.8. The Collider Principle

Which still leaves unanswered the question of what it is specifically about the Monty Hall problem that trips people up. We have identified various difficulties people have with conditional probability as an abstract concept, and we have illuminated some of the traps into which people fall. But what is it about the Monty Hall problem specifically that brings to light these inadequacies in our reasoning?

Philosopher Clark Glymour suggests a possibility in his book *The Mind's Arrows: Bayes Nets and Graphical Causal Models in Psychology* [35]. He writes,

> When two independently distributed variables, say X and Z, both influence a third variable, say Y, then *conditional on some value of Y, X and Z are not independent.* Judea Pearl gives the following illustration. Suppose that the variables are the state of the battery in your car (charged/dead), the state of the fuel tank (not empty/empty), and whether your car starts (starts/does not). Suppose that you regard the states of the battery and of the fuel tank as independent: knowing the state of the battery gives no information about the state of the fuel tank, and vice versa. Now, condition on a value of the effect—whether your car starts—by supposing that you are given the information that your car does not start. *Now* the information that your battery is charged does provide information about the state of the fuel tank.

Since the principal at work here involves two apparently independent variables metaphorically colliding with a third, this is referred to as the "collider principle." It is surely at work in the Monty Hall problem. We begin with two independent variables: our initial door choice and the location of the prize. Both have a role to play in determining the door Monty chooses to open. Consequently, the two variables should no longer be regarded as independent after we see Monty open a door.

Glymour suggests that this principle is often overlooked or misunderstood, even by people with some training in probability and statistics. It is possible that a failure to appreciate the idea is a main cause of people's difficulty with the Monty Hall problem. This suggestion was put to the test by Burns and Wieth in [12]. They noted that the psychological research suggested that the collider principle was more easily grasped in some contexts than others. They reasoned that if a failure to appreciate the principle was the heart of the matter, then translating the Monty Hall scenario into a context where the principle was more easily recognized ought to increase the rate of switching.

The minutiae of their experimental set-up and their laborious statistical analysis need not detain us. Let us proceed directly to their summary:

> Our experiments provide the first empirical evidence that at the heart of the MHD is a causal reasoning problem by supporting the five hypotheses we derived from the claim that the MHD is difficult because it requires people to understand the implications of the collider principle. Thus the results of the four experiments provide support for Glymour's (2001) speculation that the MHD is hard because it requires understanding the implications of its causal structure.

6.9. Other Work

Even in a lengthy chapter it is impossible to present more than a small portion of the work that has been done. My files contain quite a few other papers that, regrettably, I lack the space to discuss in detail.

Fox and Levav [24] fill in some more of the details regarding the subjective theorems we considered previously. They found that people's reasoning in problems of conditional probability can often be described as a three-fold process of partition-edit-count. That is, the space is first partitioned into seemingly equiprobable events in a manner suggested by the problem statement, certain events are then eliminated based on the new information, and then a solution is arrived at by counting up the number of target events among the remaining possibilities.

Krauss and Wang [52] investigated methods for making the Monty Hall problem comprehensible to people. They identified four cognitive factors that make the problem more difficult for most people, some of which we have seen previously: thinking in terms of probabilities rather than relative frequencies, forming an improper mental model of all the ways the game might play out, looking at things from the player's point of view rather than from Monty's perspective, and placing too much emphasis on the specific door Monty opened rather than on the impossibility of Monty revealing a car. In a series of clever experiments they devised versions of the problem that emphasized various combinations of these cognitive factors, with correspondingly varied levels of success in getting people to understand what was going on.

In a similar vein, other authors have investigated strategies for making the problem understandable. Have a look at Franco-Watkins, Derks, and Dougherty [26], Friedman [28], Palacios-Huerta [74], Slembeck and Tyran [84], and Tubau and Alonso [93], for example.

Finally, an amusing change of focus occurs in the work of De Beys and Verschueren [19]. Rather than study the masses of people confused by the Monty Hall problem, they investigated instead the small fraction who intuitively see their way to the correct conclusion. They suggest that such people possess a higher "working memory capacity," a factor that had previously been implicated in similar studies. I commend their paper to you for further details.

There is more besides. And all of this work sits within an even larger, and frequently contentious, literature on human probabilistic reasoning generally. Might I suggest, to any cognitive scientists who happen to be reading this, that it is time for a proper book-length treatment of the psychological literature of the Monty Hall problem?

7

Philosophical Monty

The philosophers have not been idle with regard to the Monty Hall problem. This is unsurprising. A scenario that arouses so much passion and confusion is surely something more than a mere exercise in elementary probability. The task of ferreting out what that something is has been taken up by a number of researchers.

Our primary focus will be two papers arguing that the Monty Hall problem sheds light on fundamental questions regarding the nature of probability, and the subsequent debates these papers provoked. These will form the springboard into a discussion of some subtle questions we have not had occasion to discuss previously. Whereas in the previous chapters I have primarily played the role of teacher or journalist, explaining some interesting mathematics and reporting on what various experts have said, here I intend to inject myself squarely into the debate. Reading these papers and struggling to come to terms with their clever and subtle arguments (and ultimately disagreeing with some of them) has been one of the most rewarding aspects of the research I carried out for this book. Having thought seriously about the Monty Hall problem for a number of years, it surprised me there were so many angles I had not previously considered.

You can think of this chapter as a final exam for the rest of the book, for we shall be using most of the insights we have developed to this point.

7.1. General Considerations

We begin by gathering together some useful facts.

The three-door versions of the Monty Hall problem considered to this point have shared a basic format: There are three doors, concealing two goats and one car. You select one door initially but do not open it. Monty now opens a door and it is seen to conceal a goat. You are given the opportunity to switch.

From this starting point several questions might be asked. We have focused primarily on what could be said regarding the probabilities of the two remaining doors after Monty does his thing. For example, suppose we are playing the classical version of the game and we initially chose door one. Assume Monty opens door two. We concluded that door one now has a probability of $\frac{1}{3}$ and that door three has a probability of $\frac{2}{3}$. Our decision to assign these precise probabilities was based on three major assumptions, each of them justified by our careful statement of the problem:

1. The doors were initially equiprobable.

2. Monty never opens the door you initially chose, and he reveals a goat with probability 1.

3. Monty chooses his door randomly whenever he has more than one option.

More generally, Bayes' theorem tells us that the posterior probability of a numerically specified door in light of Monty's actions is determined by three things: the prior probability assigned to that door, the probability of Monty doing what he did assuming that the car is behind that door, and the probability of Monty doing what he did minus any assumption about the location of the prize. The problem must be stated with sufficient precision to allow values to be assigned to these three quantities. If it is not, then the problem is too vague to be solved by probability theory alone.

The probabilities referred to here are *epistemic* probabilities. That is, they represent the degree of belief in a proposition warranted by the available evidence. They are to be contrasted with *statistical* probabilities, which refer to relative frequencies in long runs of trials. Epistemic probabilities depend critically on the information available to the person assigning the numbers. If two people approach the same situation with different starting information, then it is quite possible they will arrive at different epistemic probabilities. Not so with statistical probabilities. Since they are measured from hard data collected in suitable long runs of trials, they are objective and do not change from person to person.

Epistemic probabilities may be subjective in the sense that two people might assign different values if they have different information, but they are not arbitrary. That is, if two people have the same information and if this information is rich enough to permit definite values to be assigned to all of the relevant variables, then these people should agree on the correct probabilities for the situation.

There is an obvious connection between epistemic and statistical probabilities, at least in the context of the Monty Hall problem. For example, assume once more that we are playing the classical game. We initially choose door one and Monty then opens door two. We now assign an epistemic probability of $\frac{2}{3}$ to door three. We should now also believe that in a long run of trials in which we initially choose door one and Monty opens door two, the prize will be behind door three in two-thirds of the trials. This sort of probability can be referred to as *epistemic statistical* probability. It represents what we believe will happen in a long run of trials given the information at hand.

Which leads to the next point: how we should answer the question "What is the probability that we will win by switching doors?" The correct answer is not so clear, owing to the vagueness of the question. If we are thinking in terms of statistical probability, then we could interpret the question to mean, "If I play the game a large number of times and follow the strategy of switching doors every time, in what fraction of the games can I expect to win?" Let us call this the first interpretation.

The second interpretation arises in the context of a single game. Suppose you are actually playing, have initially chosen door one, and have now seen Monty open the goat-concealing door two. You are now asked for your decision. In this context, the question "What is the probability that you win by switching?" is not really well posed. The proper way to ask the question is "What is the probability that you will win by switching to door three, given what you know about Monty's procedure for opening doors, the door you chose initially, and the door Monty opened?" That is, you must condition your probability assessments on the specific information you have available to you. (As an aside, note that we are here revisiting the distinction between the conditional and unconditional forms of the problem, as discussed in section 1.11.)

The really curious thing is that these interpretations have little necessary connection to each other. The information you have might permit a conclusion about certain long-run frequencies, yet leave you in the dark as to the best course in a single play. Or you might wish to assign epistemic probabilities in a single case that do not represent your beliefs about what will happen in long runs of plays.

Let us consider some details.

7.2. Long Runs and Causal Structures

Writing in [62], philosophers Paul Moser and D. Hudson Mulder (hereafter referred to as MM) ask the following questions:

What role should statistical probability, based on a predictable distribution of outcomes in a hypothesized long run of trials, play in a decision-situation involving an individual case? Does a statistical interpretation of probability require one, in rational decision-making, to decide in an isolated individual case just as one would in a rational decision-situation involving **many repetitions of the individual case**. [Emphasis added]

The reason for the emphasis shall become clear shortly. MM then answer these questions as follows:

This paper examines these questions, arguing that it is not universally true that the rationally preferable action in a **suitable** long-run of repetitions of a particular decision-situation is likewise rationally preferable in an isolated individual case of that decision-situation. One can sometimes rationally predict the emergence of statistical correlations in long runs that are *not* indicative of a relevant causal structure operative in an isolated individual case. [Boldface added, italics in original]

And thus the gauntlet is thrown down. Note that by discussing what one can rationally predict about statistical correlations in long runs, MM are referring to epistemic statistical probabilities. We can certainly agree that raw statistical data by itself does not rationally compel you to a particular decision in a single case. Wild fluctuations observed in small numbers of single cases might average themselves out to stable relative frequencies in long runs. This, after all, is one of the main theoretical premises of statistical analysis.

That is not what MM have in mind. Instead, they envision a situation in which you have enough information to determine that a particular strategy is optimal in long runs of a particular decision situation, but nonetheless you might find it rational to follow a different strategy in a single instance of the same situation. They attempt to illustrate this point with the Monty Hall problem, which they describe as follows:

You are presented with three doors on the television show "Let's Make a Deal," and are told that there is a prize—a new car—behind one of the doors but no prize behind the other two. The game show host, Monty Hall, tells you that you will have the opportunity to pick a door to win whatever is behind it. Monty tells you that after you pick a door he will open a prizeless door from the remaining two, and that he will then give you the option of either staying with your original choice and getting a guaranteed bonus of $100 or switching to the one remaining unopened door without getting any bonus.

Let us say that you pick door number 3, and that Monty then opens the prizeless door number 2. Now you are left with the unopened doors 1 and 3. Monty offers you the opportunity either to stay with your original choice (taking whatever is behind door 3 and getting the

$100 bonus) or to switch to the single remaining unopened door (taking whatever is behind it without any bonus). Is it rational for you to stay with your original choice (door 3) rather than to switch to the other unopened door (door 1)?

Note that the first paragraph above outlines the general situation, while the second paragraph describes events occurring in a specific play of the game. This distinction shall be relevant later.

This, of course, is not the classical version of the Monty Hall problem as defined in this book. This version leaves unstated many of the key assumptions on which the familiar argument for switching depends. That is by design on MM's part, as they make clear throughout their paper and in [63]. Consequently, we should regard this as simply a new version of the problem and not one in which our previous switching argument carries any weight.

Now for the really interesting part. MM present two possible arguments for how to proceed in their version. One, which they call the "staying argument," is the familiar claim that Monty's door opening tells us only that there is no prize behind door two. Beyond that we have learned nothing about the relative likelihoods of the remaining doors, and should consequently assign a $\frac{1}{2}$ probability to each. Since we receive a $100 bonus for staying, this is the rational course to take.

Against this they present the "switching argument":

The probability that your first choice (from among the *three* doors) will win the prize is $\frac{1}{3}$. This entails that if you were to play the game 300 times, for example, your first choices in those games would probably win the prize about 100 times. If you were consistently to use the strategy of staying with your first choice every time you play, you would win about 100 prizes (plus the 300 bonuses) out of 300 games. This entails that if you were consistently using the strategy of switching in every game, you would win *all the other* games: that is, the 200 games you would have lost with the strategy of staying with your initial choice. This is because if your first choice is wrong and you switch, then you will win the prize. Hence, the probability that the prize is behind the door of your initial choice is $\frac{1}{3}$, and the probability that the prize is behind an unchosen unopened door is $\frac{2}{3}$. One might thus say (i) that the chosen door (call it C) has a $\frac{1}{3}$ probability of holding the prize, (ii) that the two doors not chosen have a $\frac{2}{3}$ combined probability of winning, and (iii) that when Monty opens a door (call it O), the $\frac{2}{3}$ probability of winning transfers to the remaining unchosen unopened door (call it U). Consequently, U gains a $\frac{2}{3}$ probability of winning. It follows, on this view, that the $\frac{2}{3}$ combined probability regarding the unchosen doors 1 and 2 transfers to door 1 individually after Monty has opened door 2, and that switching doubles your chances of winning the prize. (p. 111)

This argument begins by asserting certain statistical considerations, and proceeds to draw conclusions about rational behavior in a specific case. MM now present their main point as follows:

> This paper's thesis concerns rational decision-making under certain conditions; it does not dispute the *statistical probabilities* cited by a proponent of the Switching Argument. The thesis does, however, challenge the claim that those statistical considerations rationally *require* one to switch when given the opportunity in the game of section I (a claim made by most people who discuss the Monty Hall Problem). Clearly *if* one were to play the game described many times, and *if* one's first choice were to be correct $\frac{1}{3}$ of the time, *then* a uniform strategy of switching would produce twice as many wins as a uniform strategy of staying.

And later:

> A defender of switching will have a hard time applying the aforementioned statistical probabilities to an isolated individual case in any straightforward way.

We should note that the parenthetical statement in the first quotation above is not accurate. The specific version of the Monty Hall problem considered by MM differs in relevant ways from the standard version discussed by advocates of switching. It is true, however, that the literature abounds with sloppy and careless formulations of both the problem and its proper solution, and it is possible that is what MM had in mind.

So there it is. As MM tell the story, the information presented in their version of the Monty Hall problem permits one to rationally predict what will happen in a long run of trials, specifically that uniformly switching will lead to victories two-thirds of the time. But those statistical considerations do not transfer to an individual case. In fact, given our lack of knowledge about the method used by Monty to select his door, we are justified in assigning equal probabilities to the two remaining doors in a single play of the game. Since we receive a $100 bonus for sticking, it is reasonable to stick under these conditions. Under the precise assumptions given in MM's statement of the problem, always switching is advisable in the long run, but sticking is advisable in a single play of the game.

What do you think about that?

7.3. Switching Reconsidered

We have stressed throughout that detailed knowledge of Monty's procedure in selecting his door is essential to any solution of the problem (where by a solution to the problem we mean an ability to assign numerical probabilities to each door at each stage of the game.) In MM's version, we do not have that knowledge. Instead we can assume only that our initial choice is correct with probability $\frac{1}{3}$ and that Monty is guaranteed to open an empty door different from our choice after we make our selection. This information is plainly inadequate if our goal is to assign epistemic probabilities to the two remaining doors in an individual play of the game.

On the other hand, it is indeed adequate if our goal is to predict the long-run rate of success of switching (or sticking). The statistical correlations described in MM's switching argument are perfectly correct. Since Monty is guaranteed to open an empty door, you can lose by switching only if your initial choice conceals the car. We are stipulating that that happens in one-third of a long run of trials. It follows that switching succeeds with statistical probability $\frac{2}{3}$. So far I agree entirely with MM's view of things.

Can these statistical considerations be applied to a specific, individual case? No, they cannot. MM devote several pages to considering, and dismissing, various arguments that might be made in defense of such a move. They are working too hard, since a straightforward example should suffice to make the same point.

Let us consider two different versions of the game:

1. The classical version, in which Monty chooses randomly when he can open more than one door

2. An alternative version, in which Monty always opens the highest-numbered door available to him

In both of these versions Monty is guaranteed to reveal a goat, and we can assume that our initial choice is correct with probability $\frac{1}{3}$. Consequently, our statistical considerations apply to both cases, and we conclude that switching will win with statistical probability $\frac{2}{3}$ in the long run.

To help get our bearings, let us reconsider the statistical argument in the context of version two. Following MM, we shall assume that we play three hundred games and that we initially choose door three each time. This is an acceptable simplification, since there is nothing in our reasoning to this point that mandates the player must select his doors randomly.

Given this starting point, Monty will open door two whenever the prize is behind door one or behind door three. This will happen in two hundred cases. It is a consequence of the work we did in section 3.12 that in these two hundred cases the prize will be behind door one and behind door three with equal probability. It follows that we will win by switching in roughly one hundred of these cases, and lose in the other one hundred.

That leaves the one hundred cases where the car is behind door two. Monty will now be forced to open door one, and we will win by switching every time. We see that out of the three hundred cases, we win by switching in two hundred of them. Even when Monty follows this somewhat exotic procedure for door opening, our $\frac{2}{3}$ statistical probability is preserved.

Now return to MM's version. We have established that the statistical considerations in their switching argument hold regardless of which of the two versions we are playing. Let us then consider our options in the specific individual case described by MM. We chose door three initially and have now seen Monty open door two. If we are playing version one, then we should switch, because our available information justifies the conclusion that door one now has a probability of $\frac{2}{3}$. Things look different if we are playing version two. Now our available information tells us to assign a probability of $\frac{1}{2}$ to door one, and since we are given a bonus for sticking, that is what we should do.

It is part of MM's scenario that we have no detailed knowledge of Monty's procedure beyond the fact that he is guaranteed not to reveal the prize. So we have no basis for thinking either version one or version two is more likely to hold. Both versions lead to the same statistical probabilities for winning by switching. But they mandate different behavior in the individual case. This shows that the statistical correlations alone cannot mandate our behavior in the single case.

The point is that our long-run statistical probability of $\frac{2}{3}$ for winning by switching can come about in at least two different ways. It might, as in version one, reflect an underlying causal process that assigns a probability of $\frac{2}{3}$ to winning by switching in each individual play of the game. Alternatively, it might, as in version two, reflect the average behavior of individual trials whose probabilities are sometimes greater and sometimes smaller than $\frac{2}{3}$. In version two of the game, the unopened, unchosen door has probability $\frac{1}{2}$ with probability $\frac{2}{3}$, and has probability 1 with probability $\frac{1}{3}$. Indeed, if X denotes the probability of the remaining unopened door, then we have that the expected value of X is

$$E(X) = \left(\frac{2}{3}\right)\frac{1}{2} + \left(\frac{1}{3}\right)1 = \frac{2}{3}.$$

So I agree that in MM's version of the game, we can be confident about the statistical probability of winning by switching, and I also agree that this information is not relevant to deciding what to do in an individual case. Does this mean that I am accepting their argument that it can be rational to do in a single case what would not be rational in a long run of cases?

No, it does not.

To see why, go back to the version of the Monty Hall game provided by MM. At that time I noted there were two paragraphs to their description. In the first they outlined the general situation. In the second they described specific events taking place in an individual run of the game. The statistical correlations outlined in the switching argument applied only to the general scenario. In every play of the game one door becomes "our initial choice," while another later becomes "the unopened, unchosen door." The switching argument tells us that in long runs the first of these conceals the prize in roughly one-third of the trials, while the second conceals the prize in the remaining two-thirds.

But those are not the long runs we care about in making our decision in an individual game. For there we are given that we chose door three initially and that Monty subsequently opened door two. Consequently, the sorts of long runs relevant to making a decision in the individual case would consist entirely of situations where these two events (we choose door three and Monty opens door two) are assumed to have happened. Sadly, we are as much in the dark about the statistics of these long runs as we are about the proper epistemic probabilities to assign in the specific individual case before us.

That is why I placed in boldface the phrase "many repetitions of the individual case" in the first quotation in the previous section. It is likewise why I placed in boldface the word "suitable" in the second quotation. The individual case, as described by MM, involves Monty opening door two after we choose door three. We know nothing about what will happen in a long run of trials in which those two conditions are assumed to hold. We do know that in a long run of plays of the game, "the unopened, unchosen door" will contain the prize roughly twice as often as "our initial choice," but since these long runs will include trials in which Monty opens door one, they are not relevant to our decision in the individual game.

These sorts of considerations were not so important in the classical game. Regardless of whether we were pondering switching versus sticking in the abstract, or instead thinking about the problem in the specific context of having initially chosen door three and then seeing Monty open door two (for example), our information was rich enough to tell us all we needed to know about the resulting probabilities, whether epistemic or epistemic statistical. Not so in the MM version of the game.

There is an analogy here with the case of false positives in medical testing considered in section 1.3.2. Recall that we posited a disease that afflicts one out of every one thousand people in a population. We also had a test for the disease that never gave a false negative but did give a false positive to 5% of the people who used it. Let us suppose that Joe tests positive for the disease.

Joe is making an error if he thinks, "Nearly everyone who takes the test gets an accurate result. Therefore, it is very likely that I actually have the disease." But Joe's error is not in thinking that long-run statistical correlations (in this case represented by the vast majority who get an accurate result) necessarily determine the conclusion we should draw in an individual case. His error lies in considering the wrong kind of long run. If we are trying to comfort Joe, we would not say, "The fact that most people get an accurate result does not necessarily imply that you are sick." Instead we would say, "It is true that most people who take the test get an accurate result. But it is also true that most of the people who test positive get an inaccurate result, and those are the people to whom you should be comparing yourself."

7.4. Epistemic versus Epistemic Statistical

There is another reason to be suspicious of MM's argument. Let us accept the claims they make for their version of the Monty Hall problem. In an individual trial we should stick, but in a long run we do better to switch every time. How many times do we have to play before we stop seeing a series of individual trials and instead see a long run? If we should stick when we only play the game once, how about if we play the game twice? Three times? By their arguments we should stick if we play once but we should switch every time if we play three hundred times (their choice for a figure that represented a long run). What is the magic number where for the first time the statistical correlations kick in?

Food for thought, but this does not prove anything. And even if we reject MM's argument, we might still wonder whether their broader thesis is true. Is it possible for our epistemic probabilities in a situation to differ from our epistemic statistical probabilities regarding the same situation? Philosopher Terence Horgan took up this question in [44], written as a direct reply to MM. (See also [63] for MM's reply to that reply.)

Much of Horgan's paper was devoted to establishing the correct probabilities in the classical version of the game. Alas, this was mostly irrelevant to the arguments MM were making, since they were not considering the classical version. Horgan used the standard assumptions made in solving the problem, but MM had specifically crafted their version to make those assumptions unreasonable.

However, near the end of the paper Horgan addresses the broader question. Since I cannot improve on his discussion, I will simply reproduce it in full:

> In fact, however, single-case probabilities *cannot* ever diverge from the corresponding statistical probabilities. More precisely, the kinds of single-case and statistical probabilities that are relevant to rational decision-making—viz. *epistemic* single-case probabilities and *epistemic* statistical probabilities—cannot ever diverge from one another. This is because the kind of hypothetical long run of cases that one envisions, when one employs long-run reasoning to address single-case questions about probability, has the following two features. First, the reasoner essentially *stipulates* (perhaps implicitly) that all relevant epistemic probability-values already known are directly reflected in the statistical distributions in the envisioned long run.... Second, the reasoner also stipulates (perhaps implicitly) that all the cases in the envisioned run are relevantly similar to the reasoner's actual situation, with respect to any known information about the actual situation that is pertinent to the epistemic single-case probability-values the reasoner seeks to discover.... These two features of the envisioned hypothetical long run

of cases jointly guarantee the following form of isomorphism between the reasoner's individual situation and the envisioned run: whenever certain known epistemic single-case probabilities, together with certain pertinent available information about one's specific situation, jointly determine further epistemic single-case probabilities whose values one seeks to discover, then these latter probabilities will be identical to the corresponding epistemic *statistical* probabilities that can be ascertained in the envisioned long-run of cases.

7.5. Another Two-Player Version

Philosopher Peter Baumann, writing in [5], revived the issue in 2005. He defended MM's view, basing his argument in part on a novel, and mathematically fascinating, two-player version of the problem. We will consider that version here, saving the philosophical niceties for the next section.

Version Ten: We begin with three identical doors with one car and two goats. There are two players in the game, each of whom makes an initial door choice. Each player knows there is another player in the game, but neither player knows which door was chosen by the other player. Monty knows the door selected by each of the players and also knows the location of the car. He now opens a door according to the following procedure: If both players chose the same door, then Monty opens one of the other two doors, careful never to open the door with the car. If he has a choice of doors, then he opens a door at random. If the two players choose different doors, then Monty opens the one remaining door, regardless of what is behind it. If Monty happens to open the door with the car, then we treat it as a win for both players. Assume that you are one of the players in the game and you see Monty reveal a goat after you make your initial choice. Further assume that the other player selects his door randomly. Should you switch doors if given the opportunity to do so?

Fiendishly clever. Though care is needed, this version can be solved readily using the techniques we have already developed, as we shall show momentarily. What makes this version especially interesting is the presence of two different, and contradictory, solutions in the professional literature. Baumann presented his own solution in [5]. His solution was criticized, and a different answer offered, by Ken Levy in [53].

It is the great virtue of mathematical problems over those of philosophy that the former can generally be answered definitively while the latter frequently cannot. There is a correct answer to version ten above, and we ought to be able to come to some consensus as to what that answer is. Let us first consider the arguments of Baumann and Levy, and then attempt to resolve the matter.

Table 7.1: Possible scenarios in the two-player game
with the car assumed to be behind door one

A Chooses	B Chooses	Monty Opens	Switch	Stick
1	1	2 or 3	L	W
1	2	3	L	W
1	3	2	L	W
2	1	3	W	L
2	2	3	W	L
2	3	1	W	W
3	1	2	W	L
3	2	1	W	W
3	3	2	W	L

Baumann begins by assuming, without loss of generality, that the car is behind door one. Given this assumption, there are nine equiprobable scenarios for how the game might play out, and we record those scenarios in Table 7.1. Note that for the purposes of filling in the table, we treat the scenarios in which Monty reveals the car as a win for both strategies.

By counting up cases, we find that switching wins with probability $\frac{2}{3}$ (that is, you win in six cases out of nine by switching), while sticking wins with probability $\frac{5}{9}$. Of course, it is not a problem that these probabilities fail to add up to 1, since our idiosyncratic definitions have led to some overlap between switching victories and sticking victories. Baumann now writes,

> Consider the probabilities after Monty Hall has opened an empty door (the above nine cases thus reduce to seven). The chances of winning by switching are now $\frac{4}{7}$, whereas the chances of winning by sticking are now $\frac{3}{7}$.

Seeing Monty open an empty door eliminates the scenarios in lines six and eight in the table above. That leaves seven possibilities. Baumann treats these scenarios as equally likely, leading him to probabilities of $\frac{3}{7}$ and $\frac{4}{7}$ for sticking and switching respectively.

Ken Levy demurs at this conclusion in [53]. He offers this counterargument:

> Contrary to Baumann, the addition of a second player to the game leads the probability for both closed doors to rise to $\frac{1}{2}$. The addition of a second player changes the probabilities because it changes the initial assumptions that we may make. Consider again the one-player scenario. When A chooses door 1 and Monty Hall then opens door 3, we assume that door 3 was opened because it is part of the pair of doors that is opposed to the door that A initially chose. And because there was a $\frac{2}{3}$ probability that the car was behind one of the two doors in this pair, either door 2 or door 3, the probabilities rise to $\frac{2}{3}$. But when A competes

against B, we may no longer assume that door 3 was opened because it is part of the pair that is opposed to the door that A initially chose—i.e. doors 2 and 3. For it might just as easily have been opened because it is part of the pair of doors that is opposed to the door that B initially chose—i.e. doors 1 and 3. Since we have no reason to place door 3 in either pair, it follows that doors 1 and 2 oppose each other individually; neither opposes the other as part of a pair with door 3. Therefore the probability for either door does not rise to $\frac{2}{3}$. Instead, the probabilities for both doors rise to $\frac{1}{2}$.

So who is right? Are the probabilities after Monty opens a door $\frac{4}{7}$ and $\frac{3}{7}$, or are the two remaining doors equally likely? I shall argue that Levy's solution is incorrect, while Baumann's is correct, but incomplete.

Let us consider Levy's argument. By now we are sufficiently experienced to be suspicious of Monty Hall arguments based on intuition and hand waving. Levy's argument for the classical version does not give adequate consideration to the restrictions under which Monty labors in choosing a door to open. It is true that in choosing door one, we set up an opposition between door one, on one hand, and doors two and three, on the other. It is also true that door one has a probability of $\frac{1}{3}$ and doors two and three taken together have a probability of $\frac{2}{3}$. But it does not follow from this that when Monty eliminates door three the entire $\frac{2}{3}$ probability now shifts to door two. Further information is required to justify this conclusion. In particular, probabilities must be assigned to all of the scenarios leading Monty to open door three.

In the classical game, when I see Monty open door three after I have chosen door one, I conclude that one of two scenarios has played out:

1. Door one conceals the car; Monty chose door three randomly from the remaining two doors.

2. Door two conceals the car; Monty was forced to open door three.

The conclusion that the entire $\frac{2}{3}$ probability shifts to door two relies on the fact that scenario one happens half as often as scenario two, a reasonable proposition given the usual assumptions in the classical game. In Baumann's two-player version we should reason, after choosing door one and seeing Monty open door three, that one of three scenarios has played out:

1. Player B chose door one; door one conceals the car; Monty chose door three randomly.

2. Player B chose door one; door two conceals the car; Monty was forced to open door three.

3. Player B chose door two; Monty was forced to open door three.

Determining the probabilities of switching and sticking requires evaluating the probabilities of each of these three scenarios. Arguments about which doors are placed in opposition to one another are not adequate for that purpose.

Baumann's solution is correct (though we should note that it assumes that both players choose their doors at random, which was not stated explicitly in the problem). However, his solution is written from the perspective of one who knows the location of the car but does not know the doors chosen by the players. We might still wonder how things look to one of the players, who knows the door he chose but does not know the location of the car. We begin with the three scenarios listed above. Imagine that you are player A, that you are sitting on door one, that you have seen Monty open door three, and that door three is empty. Prior to seeing Monty open a door, you would have considered it equally likely that player B had chosen door one, door two, or door three. (Note that this assumption was built into the problem statement.) Obviously, you can now eliminate the possibility that player B chose door three. What of the other two options?

- If player B chose door two, then Monty was forced to open door three. It will conceal a goat with probability $\frac{2}{3}$.

- On the other hand, if player B chose door one then you have two further scenarios to consider.
    - One possibility is that door one conceals a car, which happens with probability $\frac{1}{2}$. In this scenario doors two and three both conceal goats. Consequently, Monty will choose his door randomly, and will select door three with probability $\frac{1}{2}$. Consequently, this scenario occurs with probability $\frac{1}{6}$.
    - The alternative is that player B chose door one and it conceals a goat, while door two conceals the car. Now Monty is forced to open door three. This happens with probability $\frac{1}{3}$, since that is the probability of finding the car behind door two.

    It follows that the probability that B chose door one given that Monty opened door three and showed that it was empty is

$$\frac{1}{3} + \frac{1}{6} = \frac{1}{2}.$$

Note that this is smaller than $\frac{2}{3}$. You are more likely to observe Monty open an empty door three in those cases where B chose door two than where B chose door one. It follows that the scenarios in which B chose door two should collectively be assigned a higher probability than the scenarios in which B chose door one.

We can go further by invoking the proportionality principle (see section 3.12). Initially we thought B choosing door one had the same probability as B choosing door two. Then an event occurred that happens with probability $\frac{1}{2}$ under the assumption that B chose door one, and with probability $\frac{2}{3}$ under the assumption that B chose door two. It follows that the updated probabilities for these two possibilities must preserve this ratio. That is, if B_1 and B_2 are respectively the events that B chose door one or door two, then I should have

$$\frac{P(B_1)}{P(B_2)} = \frac{1/2}{2/3} = \frac{3}{4}.$$

Thus, the scenarios in which B chose door one should collectively be given a probability of $\frac{3}{7}$, while the remaining possibilities should collectively be given a probability of $\frac{4}{7}$ (since the probabilities must add up to 1).

Since this leaves unresolved the question of whether A ought to switch or stick, we must persist. From A's perspective, there are now four scenarios remaining. We denote these scenarios by ordered triples of the form

(A's Choice, B's Choice,   Car Location).

The two scenarios in which B chose door two are now written $(1, 2, 2)$ and $(1, 2, 1)$. The information at hand gives no basis for thinking the car is more likely to be behind one of the doors than the other. Since both of these together should have a probability of $\frac{4}{7}$, we find that each scenario individually has probability $\frac{2}{7}$.

The remaining two scenarios are $(1, 1, 2)$ and $(1, 1, 1)$, and it is here that Baumann's problem offers up its final subtlety. These two scenarios are not equally likely, you see. If both players choose door one and the car is behind door two, then it is certain that Monty will open door three. But if both players choose door one and the car is behind door one, then Monty only chooses door three with probability $\frac{1}{2}$.

Once more invoking the proportionality principle leads to the conclusion that $(1, 1, 2)$ should be viewed as twice as likely as $(1, 1, 1)$. And since both together have a probability of $\frac{3}{7}$, we should find that $(1, 1, 2)$ has probability $\frac{2}{7}$, while $(1, 1, 1)$ has probability $\frac{1}{7}$.

The scenarios in which you win by switching are $(1, 2, 2)$ and $(1, 1, 2)$. Therefore, you win by switching with probability $\frac{2}{7} + \frac{2}{7} = \frac{4}{7}$. Winning by sticking has a probability of $\frac{3}{7}$.

Tricky stuff. The conclusion, however, is certainly correct. I know this because a colleague of mine ran a Monte Carlo simulation for me with fifty thousand trials, and the empirical probabilities worked out to be satisfyingly close to the theoretical values derived here.

Of course, this problem could have been solved more directly using Bayes' theorem or by working out the tree diagram for the problem. I find the method above, with its double application of the proportionality principle, to be more elegant.

Table 7.2: Probabilities assigned by the players to different propositions in the two-player game

	Other Door Wins	Door Two Wins
A's Probability	4/7	4/7
B's Probability	4/7	3/7

7.6. Single-Case Probabilities

That was mere appetizer. The main course is Baumann's conclusion.

Let us first agree to the following principle: If two rational people have precisely the same relevant information, and if that information fully determines their rational degree of belief in a proposition, then both people should assign the same probability to that proposition. Baumann refers to this as a principle of non-arbitrariness, and he is unlikely to find anyone demurring from it.

Baumann claims that this principle is violated in the case of his two-player Monty Hall problem. Assume for the moment that player A has initially chosen door one and that player B has initially chosen door two. Monty then opens door three, and it turns out to be empty. How will the situation now look to each of the two players?

Following the logic from the previous section, player A will now believe that door two has probability $\frac{4}{7}$, while door one has probability $\frac{3}{7}$. Player B will think exactly the reverse, that door one has probability $\frac{3}{7}$, while door two has probability $\frac{4}{7}$. We record this information in Table 7.2 (following the example of Jan Sprenger in [87]).

Is there anything remarkable in this?

Perhaps so. Look again at that principle of non-arbitrariness. We are certainly covered on the second part of it. In the two-player Monty Hall problem, the probabilities the players ought to assign to each of the doors is fully determined by the available information. But do the players possess the same relevant information? The only conceivable difference lies in the initial choices made by the players. Player A knows he chose door one but does not know which door was chosen by player B. On the other hand, player B knows he chose door two but does not know which door A chose.

Baumann argues that this difference is not relevant. I will turn the floor over to him:

> In the modified Monty Hall case the two players have different information but... it is hard to see how that could be relevant with respect to the relevant question: which door—door one or door two—will have what probability of being the winner in this particular game....

There is a very tempting objection to this last step which in the end turns out to be subtly misleading. Suppose A has originally picked door one and he knows this of course. This information plus the information that Monty has opened door three plus some probabilistic reasoning about switching to the "other" door seems to suggest that he should make door two his final choice. He couldn't arrive at this conclusion without the information that he has originally picked door one. Hence that information seems clearly relevant. Furthermore, it is information B (who has originally picked door two) does not have....

However... an expression like "the probability that the other door will win" and an expression like "the probability that door two will win" do have different meanings, and this difference matters here. All the relevant information the chooser has can be expressed in terms of "the original door" and "the other door." There is no relevant information having to do with the fact that one door is, say, door one. Hence A and B share the same relevant information.

If Baumann is right, then we do indeed have a paradox. If A and B have the same information regarding the possible locations of the prize, then they really ought to make the same probability assignments. That they do not do so in this case might indicate something fundamentally illegitimate about assigning single-case probabilities.

I shall argue, however, that Baumann is not right. The problem begins with the table above. About what, exactly, do A and B disagree? Not about the probability of winning by switching. They agree that by switching they will move from a door with probability $\frac{3}{7}$ to a door of probability $\frac{4}{7}$. Their information regarding the benefits of switching as a strategy is identical, and they rightly arrive at the same probability assignment.

The disagreement is over the probability to assign to door two (and door one) given the manner in which this particular play of the game has unfolded. Their opinions differ only with regard to the conditional probabilities to assign given the specific history of this one game. With respect to *that* question, the players have very different information indeed. The reason is one we have had cause to point out many times in this book: any restriction placed on Monty in choosing his door is relevant to determining the posterior probabilities after Monty opens his door. It follows that if the players have relevantly different information regarding Monty's limitations, it is not unreasonable for them to make different probability assignments.

When player A sees Monty open door three, he knows Monty did that in part because he was forbidden by the rules from opening door one. Player B does not know that. This difference will affect the way they evaluate their probabilities, as will the fact that B knows Monty was not allowed to open door two (which A does not know). It is not surprising that they assign different probabilities to door two, because the information they have with regard to Monty's actions is not the same.

Baumann also encourages us to consider the following thought experiment:

Suppose A and B both forget what their initial pick was. However, both know that if they answer "Switch!" (or "Stick!" for that matter) when asked for their final answer the host will follow instructions and, knowing the contestants' initial picks, give them what is behind the respective (other or the same) door. It seems obvious that this variation does not introduce any relevant differences into the game. However, the difference in A's and B's information concerning their initial picks drops out of the picture. Hence, the information about initial picks cannot be relevant. In other words, it does not seem to make any difference for one's final choice whether one remembers which particular door one had picked initially. Hence, that information is not relevant.

If the question is whether to switch or stick when given the opportunity, then this paragraph is certainly correct. The initial picks do not affect the judgment that switching is the wisest strategy to follow, both in the individual case and in the long run. Recall, however, that A and B do not disagree on this point. The disagreement is over the posterior probabilities that attach to particular doors, and something critical to that issue has, indeed, been lost when the players forget their initial picks. In the given example where A and B remember their choices we saw that A assigned a probability of $\frac{4}{7}$ to door two, while B assigned a probability of $\frac{3}{7}$ to the same door. It was these probability assignments that were said to lead to a paradox. But if A and B forget their initial choices, then they can no longer make these probability assignments. They can still conclude that switching is the way to go, but they never disagreed on that point.

There is no way around it. In the specific game described, the players have different relevant information regarding the location of the prize after Monty opens his door. Consequently, there is nothing paradoxical in their differing posterior probabilities.

7.7. Counterfactuals and Classical Monty

Let us go one more round with Baumann, this time over the question of the familiar one-player game. It only seems reasonable that we close the book where we began, and ponder once more the seemingly endless mysteries of the classical Monty Hall problem.

The symbol "(N)" in this quotation refers to the principle of non-arbitrariness discussed earlier.

If A has originally picked door 1, and Monty Hall has just opened door 3, then the probability that the originally unchosen door ("the other door,") will win is, of course, $\frac{2}{3}$. Since A knows the other door = door 2, he will update the probability that door 2 will win to $\frac{2}{3}$; accordingly, the probability that door 1 will win would be $\frac{1}{3}$. In other words, he knows that the probability that door 2 (or door 1 for that matter) will win is either $\frac{2}{3}$ or $\frac{1}{3}$; what it is depends only on his initial pick. If door 2 has a $\frac{2}{3}$ chance of winning, that is because A had (given the circumstances of the case) initially picked door 2, then the probabilities would be the reverse ones. But all that just seems absurd. It violates (N) and its relatives. Probabilities do not vary with irrelevant factors. Now, whether one has originally picked one or the other door, cannot make a difference as to one's (epistemic) probabilities. There is no relevant difference between the two doors (see above). Hence, there is something wrong with the conclusion that if A has originally picked door 1, then door 2 will have a $\frac{2}{3}$ winning chance whereas if A has originally picked door 2, door 1 will have a $\frac{2}{3}$ winning chance. **The actual case in which A originally picks door 1 doesn't differ from the counterfactual case in which A originally picks door 2 in such a way that the distribution of probabilities in both cases should differ.** (Emphasis added)

There is much to discuss here, and that boldface sentence provides a welcome entry point. Has Baumann placed his finger on something paradoxical? I am afraid he has not. The boldface statement, you see, is not correct. It overlooks the fact that *Monty's choice of door to open is constrained by the player's initial choice.*

That is why Baumann's counterfactual argument does not work. You cannot watch Monty open door three and then retroactively pretend that you initially chose door two instead of door one. If you are going to mentally alter your initial behavior, you must also give Monty the opportunity to alter his own. Monty's decision to open door three was based in part on the fact that door one was unavailable to him, owing to the player's initial choice. Had the player chosen door two, Monty would not have been laboring under that restriction. (He would have a different restriction in its place.) For that reason, the counterfactual case differs from the actual case in a way that is relevant for assigning probabilities.

With that in mind, we can spot some errors in the remainder of this paragraph. In the specific case where A picks door one and Monty then opens door three, the decision by A to update his probabilities to $\frac{1}{3}$ and $\frac{2}{3}$ for doors one and two respectively is not based on any considerations about which is "the other door." It is based instead on a probability calculation conditioned on the specific information available in that play of the game. That includes A's initial choice, the door Monty opened, and the precise manner in which Monty selects his door. It is a quirk about the classical game that the long-run statistical probability of "winning by switching" is equal to the probability that should be assigned to "the other door" in any individual play, but the former is not used to justify the latter.

This is precisely the point we raised at the start of the chapter. When you are in the middle of the game and have to make an actual decision, it is no longer a well-posed question to ask, "What is the probability of winning by switching?" The question must be, "What is the probability of winning by switching *given* the information I have specific to this play of the game?" It is likewise incorrect to say the probability assigned to doors one and two depends only on the player's initial pick (which would, indeed, make the single-case probability assignments seem a bit arbitrary). In reality it depends both on the initial pick *and on Monty's subsequent actions.* And since what Monty does depends in part on what you do initially, the two cannot be considered in isolation from one another.

7.8. Concluding Thoughts

If there is something about the Monty Hall problem that sheds light on deeper issues regarding the nature of probability, I do not believe these papers have revealed it. That notwithstanding, these authors certainly provide ample food for thought.

There is much more in these papers than what I have discussed here. I invite you to peruse the entire collection. Start with Moser and Mulder's original paper [62], the reply by Horgan [44], and the retort by Moser and Mulder [63]. Then move on to Baumann's paper [5], Levy's reply [53], Baumann's subsequent rebuttal of Levy [6], and most recently, to Sprenger's criticism of Baumann [87]. The true Monty Hall connoisseur can do no less!

8

Final Monty

8.1. Monty Strikes Again!

I finished the first draft of this manuscript on April 7, 2008, almost a year after I began. Felt real good about myself. Strode out of my office, rubbed my weary eyes, and smiled, satisfied that at long last I had lifted the Monty Hall monkey from my aching back.

It did not last long. The following day I showed up to my office and began my morning routine. Part of this routine is a perusal of that day's *New York Times*. Lurking in the Science section was this headline: "And Behind Door No. 1, a Fatal Flaw." The first paragraph of the accompanying article [90]:

> The Monty Hall Problem has struck again, and this time it's not merely embarrassing mathematicians. If the calculations of a Yale economist are correct, there's a sneaky logical fallacy in some of the most famous experiments in psychology.

Every time I think I'm out they keep pulling me back in!

The Yale economist in question is M. Keith Chen. His criticism is directed at research into so-called choice-induced dissonance. According to the research, people tend to rationalize past decisions by elevating their opinion of the choice made and downgrading their opinion of the choice rejected. As a trivial example, suppose we offer a person a choice between a red candy and a blue candy, having previously ascertained that he has no preference between them. Suppose the person chooses red. If we later ask him to once more rank his preferences between the two candies, he is likely to describe a clear preference for red over blue. The fact of having chosen red seems to affect his assessment of the merits of red over blue.

It is Chen's claim that a methodological flaw afflicts all of the numerous studies of this phenomenon. For example, one experimental setup for testing this phenomenon begins with people ranking their preferences for certain goods on a numerical scale. The experimenter then selects three goods, which we shall denote by A, B, and C, that are ranked equally by the subject. I will let Chen himself continue the story [13]:

> In a second stage then, a subject is asked to choose between a *randomly* chosen two of these items, say A and B. Calling the object which the subject chooses A, the subject is then asked to choose between B (the initially rejected item), and C (the third item that was rated [equal to A and B]. If subjects are more likely to choose C then B in this choice, they are said to suffer from CD [cognitive dissonance].
>
> I argue that this was to be expected in subjects with no CD. In fact, subjects should be *expected* to choose good C 66% of the time.

The premise here is that people are never truly indifferent between options. Subtle preferences among A, B and C can be obscured by the discreteness of the measuring scale used (so two options might both receive a ranking of 4 out of 5, even though one of those options is really a 3.9 while the other is a 4.1). If that is the case, then the subject's choice of A over B reflects a subtle preference for A.

If you think this is irrelevant, that knowledge of the subject's preference for A over B tells us nothing of his preferences between B and C, then I urge you to think again. There are only three possible orderings for the subject's preferences given that he prefers A to B. Listing them from most favorite to least favorite, we have

$$ABC \quad CAB \quad ACB.$$

In two out of three cases the subject also prefers C to B. Consequently, statistics alone are sufficient to explain the data. No highfalutin notions about cognitive dissonance are needed.

I will leave it to the experts in the field to determine if Chen's criticisms have any merit. But given the track record of researchers in handling Monty Hall type reasoning, I would take his charges very seriously indeed.

8.2. Sequel Stuffers

When I first began telling people I wanted to write a book on the Monty Hall problem, I encountered quite a lot of incredulity. "A whole book?" they would ask. "On the Monty Hall problem? What is there to say beyond, 'You should switch'?"

So it is with a certain satisfaction that I note that I could easily write a second book with all the material I have left out of this one. As mentioned in Chapter 6, the psychological literature alone could be the focus of a lengthy treatise. Mathematically, there are variations that lead naturally to discussions of concepts in game theory [2; 25], and other variations best approached with sophisticated techniques from statistical decision theory [14]. Philosophers have said a lot more about the problem than what I included in Chapter 7, and I would call your attention to [11; 15; 96; 97; 100] for just a few that I found especially intriguing.

I have computer scientists using the problem to shed light on certain esoteric questions in logic [50]. Law professors using it to discuss the difficulties juries have in assessing probabilistic evidence [99]. Geneticists using it in discussions of the importance of Bayes' theorem for doctors [70]. Educators applying it to numerous pedagogical ends [49; 71].

Get the idea?

It would seem the fascinations of the Monty Hall problem are endless. My book, however, is not. Thank you for your attention, and good night.

Appendix: A Gallery of Variations

I have occasionally found it convenient to present certain Monty Hall variations in italics, to set them off from the rest of the text. Other variations did not merit this treatment, usually because they were only slightly different from the italicized problems. In addition to the variations considered here, the literature records several others that I have not included in this book. For convenience, I have included in this Appendix a complete list of Monty Hall variations discussed in the text. Since I am certain the basic setup is now familiar to everyone, I will list these variations in abbreviated form. You can assume the doors are identical unless I specifically say otherwise, and that Monty will never open (or point to) the door you initially chose. In the three-door versions you may also assume there is one car and two goats, again unless I specifically say otherwise.

1. **Section 2.1.** Three doors; Monty always reveals a goat and chooses randomly when he has a choice.

2. **Section 3.1.** Three doors; Monty opens a door at random.

3. **Section 3.9.** Three doors; Monty points to a door and asserts that it conceals a goat. His assertions are correct with probability $p \geq \frac{2}{3}$.

4. **Section 3.10.** Three doors; Monty always reveals a goat and chooses randomly when he has a choice. The car is initially placed behind doors one, two, and three with probabilities p_1, p_2, and p_3 respectively.

5. **Section 3.12.** Three doors; Monty always reveals a goat. He opens the lowest-numbered door available to him with probability p and the highest-numbered door with probability $1 - p$.

6. **Section 4.1.** There are n doors with $n \geq 3$. They conceal one car and $n - 1$ goats. You make an initial choice. Monty now reveals a goat, choosing randomly from among his options. You are then given the option of switching. After making your choice, Monty again reveals a goat, still choosing randomly from among his options. You are again given the option of switching. This continues until only two doors remain.

7. **Section 5.1.** There are three doors, concealing three different prizes ranked unambiguously as Bad, Middle, and Good. Benevolent Monty always reveals the lowest-valued prize concealed by a door different from the one you chose. Malevolent Monty always reveals the highest-valued prize. Random Monty opens a door randomly.

8. **Section 5.2.** There are three doors and two players. Player one chooses a door, and then player two chooses a different door. If both players choose goats, then one is eliminated at random. If one chose the car, then the other is eliminated. Monty then opens the door chosen by the eliminated player.

9. **Section 5.3.** There are three doors. There are also two hosts. One chooses his door randomly when given a choice of doors; the other always opens the highest-numbered door. The presiding host is chosen by a coin toss. You initially choose door one and then see Monty open door three.

10. **Section 5.4.** There are n doors with $n \geq 3$. There are j cars and $n - j$ goats. Monty opens a door at random.

11. **Section 5.4.** There are n doors with $n \geq 3$. There are j cars and $n - j$ goats. Monty tells us he will reveal a goat with probability p and a car with probability $1 - p$. We must decide ahead of time whether we will switch or stick after his revelation.

12. **Section 5.4.** There are n doors with $n \geq 3$. There are j cars and $n - j$ goats. Monty opens m doors, with $1 \leq m \leq n - 2$, revealing k cars and $m - k$ goats.

13. **Section 5.5.** There are n doors with $n \geq 3$. There are m types of prize with values $v_1, v_2, \ldots, v_m$. Monty opens a door at random.

14. **Section 7.2.** There are three equiprobable doors. Monty always reveals a goat. We get a \$100 bonus for sticking. This is all we know.

15. **Section 7.5.** There are three doors. There are also two players. Each player chooses a door without knowing the door chosen by the other player. Monty knows both choices. If the players chose different doors, then Monty opens the one remaining door. If the players chose the same door, then Monty reveals a goat, choosing randomly when he has more than one option.

Bibliography

[1]   Eric Aaron, Michael Spivey-Knowlton, "Frequency vs. Probability Formats: Framing the Three Doors Problem," in *Proceedings of the Twentieth Annual Conference of the Cognitive Science Society*, Lawrence Erlbaum Associates, 1998, pp. 13–19.

[2]   Herb Bailey, "Monty Hall Uses a Mixed Strategy," *Mathematics Magazine*, Vol. 73, No. 2, April 2000, pp. 135–41.

[3]   Ed Barbeau, "The Problem of the Car and Goats," *College Mathematics Journal*, Vol. 24, No. 2, Mar. 1993, pp. 149–54.

[4]   Maya Bar-Hillel, Ruma Falk, "Some Teasers Concerning Conditional Probabilities," *Cognition*, Vol. 11, No. 2, 1982, pp. 109–22.

[5]   Peter Baumann, "Three Doors, Two Players, and Single-Case Probabilities," *American Philosophical Quarterly*, Vol. 42, No. 1, January 2005, pp. 71–9.

[6]   Peter Baumann, "Single-Case Probabilities and the Case of Monty Hall: Levy's View," *Synthese*, Vol. 162, No. 2, May 2008, pp. 265–73.

[7]   Richard Bedient, "The Prisoner's Paradox Revisited," *American Mathematical Monthly*, Vol. 101, No. 3, March 1994, p. 249.

[8]   Jakob Bernoulli, *The Art of Conjecturing*, translated by Edith Dudley Sylla, Johns Hopkins University Press, Baltimore, 2006.

[9]   Michael Birnbaum. "Base Rates in Bayesian Inference," Chapter Two in *Cognitive Illusions*, R. F. Pohl, ed., Psychology Press, New York, 2004, pp. 43–60.

[10] Alan Bohl, Matthew Liberatore, Robert Nydick, "A Tale of Two Goats... and a Car, or the Importance of Assumptions in Problem Solutions," *Journal of Recreational Mathematics*, Vol. 27, No. 1, 1995, pp. 1–9.

[11] Darren Bradley, Branden Fitelson, "Monty Hall, Doomsday and Confirmation," *Analysis*, Vol. 63, No. 1, January 2003, pp. 23–31.

[12] Bruce Burns, Mareike Wieth, "The Collider Principle in Causal Reasoning: Why the Monty Hall Dilemma Is So Hard," *Journal of Experimental Psychology: General*, Vol. 133, No. 3, 2004, pp. 436–49.

[13] M. Keith Chen, "Rationalization and Cognitive Dissonance: Do Choices Affect or Reflect Preferences?" unfinished working paper, 2008.

[14] Young Chun, "On the Information Economics Approach to the Generalized Game Show Problem," *American Statistician*, Vol. 53, No. 1, February 1999, pp. 43–51.

[15] Charles Cross, "A Characterization of Imaging in Terms of Popper Functions," *Philosophy of Science*, Vol. 67, No. 2, June 2000, pp. 316–38.

[16] G. M. D'Arioano, R. D. Gill, M. Keyl, R. F. Werner, B. Kummerer, H. Maassen, "The Quantum Monty Hall Problem," *Quantum Information and Computation*, Vol. 2, 2002, pp. 355–66.

[17] F. N. David, *Games, Gods and Gambling*, Dover Publications, New York, 1998.

[18] Richard Dawkins, "In Defence of Selfish Genes," *Philosophy*, Vol. 56, 1981, pp. 556–73.

[19] Wim De Neys, Niki Verschueren, "Working Memory Capacity and a Notorious Brain Teaser: The Case of the Monty Hall Dilemma," *Experimental Psychology*, Vol. 53, No. 2, 2006, pp. 123–31.

[20] P. Diaconis, S. Zabell, "Some Alternatives to Bayes's Rule," in *Information Pooling and Group Decision Making: Procedings of the Second University of California, Irvine Conference on Political Economy*, B. Grofman and G. Owen, eds., JAI Press, Greenwich, 1986, pp. 25–38.

[21] E. Engel, A. Ventoulias, "Monty Hall's Probability Puzzle," *Chance*, Vol. 4, No. 2, 1991, pp. 6–9.

[22] Ruma Falk, "A Closer Look at the Probabilities of the Notorious Three Prisoners," *Cognition*, Vol. 43, 1992, pp. 197–223.

[23] Nicholas Falletta, *The Paradoxicon*, John Wiley and Sons, New York, 1983.

[24] Craig Fox, Jonathan Levav, "Partition-Edit-Count: Naive Extensional Reasoning in Judgment of Conditional Probability," *Journal of Experimental Psychology: General*, Vol. 133, No. 4, 2004, pp. 626–42.

[25] Luis Fernandez, Robert Pinon, "Should She Switch? A Game-Theoretic Analysis of the Monty Hall Problem," *Mathematics Magazine*, Vol. 72, No. 3, June 1999, pp. 214–17.

[26] A. M. Franco-Watkins, P. L. Derks, M. R. P. Dougherty, "Reasoning in the Monty Hall Problem: Examining Choice Behaviour and Probability Judgements," *Thinking and Reasoning*, Vol. 9, No. 1, 2003, pp. 67–90.

[27] John E. Freund, "Puzzle or Paradox?" *American Statistician*, Vol. 19, No. 4, Oct. 1965, pp. 29, 44.

[28] D. Friedman, "Monty Hall's Three Doors: Construction and Deconstruction of a Choice Anomaly," *The American Economic Review*, Vol. 88, No. 4, September 1998, pp. 933–46.

[29] Martin Gardner, "Problems Involving Questions of Probability and Ambiguity," *Scientific American*, Vol. 201, No. 4, April 1959, pp. 174–82.

[30] Martin Gardner, "How Three Modern Mathematicians Disproved a Celebrated Conjecture of Leonhard Euler," *Scientific American*, Vol. 201, No. 5, May 1959, p. 188.

[31] J. P. Georges, T. V. Craine, "Generalizing Monty's Dilemma," *Quantum*, March/April 1995, pp. 17–21.

[32] G. Gigerenzer, R. Hertwig, E. van den Broek, B. Fasolo, K. Katsikopoulos, "A Thirty Percent Chance of Rain Tomorrow: How Does the Public Understand Probabilistic Weather Forecasts?" *Risk Analysis*, Vol. 25, No. 3, 2005, pp. 623–29.

[33] Leonard Gillman, "The Car and the Goats," *American Mathematical Monthly*, Vol. 99, No. 1, January 1992, pp. 3–7.

[34] Thomas Gilovich, Victoria Husted Medvec, Serena Chen, "Commission, Omission, and Dissonance Reduction: Coping with Regret in the 'Monty Hall' Problem," *Personality and Social Psychology Bulletin*, Vol. 21, No. 2, February 1995, pp. 182–90.

[35] Clark Glymour, *The Mind's Arrows: Bayes Nets and Graphical Causal Models in Psychology*, MIT Press, Cambridge, 2001.

[36] Donald Granberg, "Cross-Cultural Comparison of Responses to the Monty Hall Dilemma," *Social Behavior and Personality*, Vol. 27, No. 4, 1999, pp. 431–48.

[37] Donald Granberg, "A New Version of the Monty Hall Dilemma with Unequal Probabilities," *Behavioural Processes*, Vol. 48, 1999, pp. 25–34.

[38] Donald Granberg, Nancy Dorr, "Further Exploration of Two-Stage Docision Making in the Monty Hall Dilemma," *American Journal of Psychology*, Vol. 111, No. 4, 1998, pp. 561–79.

[39] Donald Granberg, Thad A. Brown, "The Monty Hall Dilemma," *Personality and Social Psychology Bulletin*, Vol. 21, No. 7, July 1995, pp. 711–23.

[40] Ian Hacking, *The Emergence of Probability*, 2nd ed., Cambridge University Press, New York, 2006.

[41] Mark Haddon, *The Curious Incident of the Dog in the Night-Time*, Vintage Contemporaries, New York, 2003.

[42] J. M. Hammersley, D. C. Handscomb, *Monte Carlo Methods*, Methuen, London, 1964.

[43] Paul Hoffman, *The Man Who Loved Only Numbers*, Hyperion, New York, 1998.

[44] Terence Horgan, "Let's Make a Deal," *Philosophical Papers*, Vol. 26, No. 3, 1995, pp. 209–22.

[45] J. Howard, C. Lambdin, D. Datteri, "Let's Make a Deal: Quality and Availability of Second-Stage Information as a Catalyst for Change," *Thinking and Reasoning*, Vol. 13, No. 3, 2007, pp. 248–72.

[46] S. Ichikawa, H. Takeichi, "Erroneous Beliefs in Estimating Posterior Probability," *Behaviormetrika*, Vol. 27, 1990, pp. 59–73.

[47] Richard Isaac, *The Pleasures of Probability*, Springer-Verlag, New York, 1995.

[48] Brian Kluger, Steve Wyatt, "Are Judgment Errors Reflected in Market Prices and Allocations? Experimental Evidence Based on the Monty Hall Problem," *Journal of Finance*, Vol. 59, No. 3, June 2004, pp. 969–97.

[49] M. Kelley, "Demonstrating the Monty Hall Dilemma," *Teaching of Psychology*, Vol. 31, No. 3, 2004, pp. 193–95.

[50] Barteld Kooi, "Probabilistic Dynamic Epistemic Logic," *Journal of Logic, Language and Information*, Vol. 12, 2003, pp. 381–408.

[51] Steven Krantz, *Techniques of Problem Solving*, American Mathematical Society, Providence, 1997.

[52] Stefan Krauss, X. T. Wang, "The Psychology of the Monty Hall Problem: Discovering Psychological Mechanisms for Solving a Tenacious Brain

Teaser," *Journal of Experimental Psychology: General*, Vol. 13, No. 1, pp. 3–22.

[53] Ken Levy, "Baumann on the Monty Hall Problem and Single-Case Probabilities," *Synthese*, Vol. 158, No. 1, 2007, pp. 139–51.

[54] Anthony Lo Bello, "Ask Marilyn: The Mathematical Controversy in Parade Magazine," *Mathematical Gazette*, Vol. 75, No. 473, October 1991, pp. 275–77.

[55] Stephen Lucas, Jason Rosenhouse, "Optimal Strategies in the Progressive Monty Hall Problem," forthcoming in *Mathematical Gazette*.

[56] Sidney Luckenbach, *Probabilities, Problems and Paradoxes: Readings in Inductive Logic*, Dickenson, Encino, 1972.

[57] Phil Martin, "The Monty Hall Trap," in *For Experts Only*, P. Granovetter and E. Granovetter, eds., Granovetter Books, Cleveland Heights, 1993.

[58] Robert M. Martin, *There Are Two Errors in the the Title of This Book*, Broadview Press, Ontario, 2002.

[59] E. H. McKinney, "Generalized Birthday Problem," *American Mathematical Monthly*, Vol. 73, No. 4, 1966, pp. 385–87.

[60] J. P. Morgan, N. R. Chaganty, R. C. Dahiya, M. J. Doviak, "Let's Make a Deal: The Player's Dilemma," *American Statistician*, Vol. 45, No. 4, November 1991, pp. 284–87.

[61] J. P. Morgan, N. R. Chaganty, R. C. Dahiya, M. J. Doviak, "Let's Make a Deal: The Player's Dilemma [Rejoinder]," *American Statistician*, Vol. 45, No. 4, November 1991, p. 289.

[62] Paul K. Moser, D. Hudson Mulder. "Probability in Rational Decision-Making," *Philosophical Papers*, Vol. 23, No. 2, 1994, pp. 109–28.

[63] Paul K. Moser, D. Hudson Mulder, "Probability and Evidence: A Response to Horgan," unpublished.

[64] F. Mosteller, *Fifty Challenging Problems in Probability, with Solutions*, Addison-Wesley Reading, 1965.

[65] Paul J. Nahin, *Duelling Idiots and Other Probability Puzzlers*, Princeton University Press, Princeton, 2000.

[66] Barry Nalebuff, "Choose a Curtain, Duel-ity, Two Point Conversions and More," *Economic Perspectives*, Puzzles Column, Vol. 1, No. 1, Fall 1987, pp. 157–63.

[67] Raymond Nickerson, "Ambiguities and Unstated Assumptions in Probabilistic Reasoning," *Psychological Bulletin*, Vol. 120, No. 3, 1996, pp. 410–33.

[68] Martine Nida-Rumelin, "Probability and Direct Reference, Three Puzzles of Probability Theory: The Problem of the Two Boys, Freund's Problem and the Problem of the Three Prisoners," *Erkenntnis*, Vol. 39, No. 1, July 1993, pp. 51–78.

[69] Eugene Northrop, *Riddles in Mathematics: A Book of Paradoxes*, D. Van Nostrand Company, Princeton, 1944, pp. 173–76.

[70] Konrad Oexle, "Useful Probability Considerations in Genetics: The Goat Problem With Tigers and Other Applications of Bayes' Theorem," *European Journal of Pediatrics*, Vol. 165, 2006, pp. 299–305.

[71] Ken Overway, "Empirical Evidence or Intuition: An Activity Involving the Scientific Method," *Journal of Chemical Education*, Vol. 84, No. 4, April 2007, pp. 606–8.

[72] Scott E. Page, "Let's Make a Deal," *Economics Letters*, Vol. 61, 1998, pp. 175–80.

[73] Massimo Piattelli-Palmarini, "Probability Blindness: Neither Rational Nor Capricious," *Bostonia*, March/April 1991, pp. 28–35.

[74] I. Palacios-Huerta, "Learning to Open Monty Hall's Doors," *Experimental Economics*, Vol. 6, 2003, pp. 235–51.

[75] J. Paradis, P. Viader, L. Bibiloni, "A Mathematical Excursion: From the Three-door Problem to a Cantor-Type Set," *American Mathematical Monthly*, Vol. 106, No. 3, March 1999, pp. 241–51.

[76] B. D. Puza, D. G. W. Pitt, T. J. O'Neill, "The Monty Hall Three Doors Problem." *Teaching Statistics*, Vol. 27, No. 1, Spring 2005, pp. 11–15.

[77] V. V. Bapeswara Rao and M. Bhaskara Rao, "A Three-Door Game Show and Some of its Variants," *The Mathematical Scientist*, Vol. 17, No. 2, 1992, pp. 89–94.

[78] Jeffrey Rosenthal, "Monty Hall, Monty Fall, Monty Crawl," *Math Horizons*, September 2008, pp. 5–7.

[79] Jeffrey Rosenthal, *Struck by Lightning: The Curious World of Probabilities*, Joseph Henry Press, Washington, D.C., 2006.

[80] Steve Selvin, "A Problem in Probability" (letter to the editor), *American Statistician*, Vol. 29, No. 1, 1975, p. 67.

[81] Steve Selvin, "On the Monty Hall Problem" (letter to the editor), *American Statistician*, Vol. 29, No. 3, 1975, p. 134.

[82] J. Michael Shaughnessy, Thomas Dick, "Monty's Dilemma: Should You Stick or Switch?" *Mathematics Teacher*, Vol. 84, 191, pp. 252–56.

[83] S. Shimojo, S. Ichikawa, "Intuitive Reasoning About Probability: Theoretical and Experiential Analysis of Three Prisoners," *Cognition*, Vol. 32, 1989, pp. 1–24.

[84] Tilman Slembeck, Jean-Robert Tyran, "Do Institutions Promote Rationality? An Experimental Study of the Three-Door Anomaly," *Journal of Economic Behavior and Organization*, Vol. 54, 2004, pp. 337–50.

[85] John Maynard Smith, *Mathematical Ideas in Biology*, Cambridge University Press, London, 1968.

[86] Raymond Smullyan, *The Riddle of Scheherazade and Other Amazing Puzzles*, Harcourt, San Diego, 1997.

[87] Jan Sprenger, "Probability, Rational Single-Case Decisions and the Monty Hall Problem," forthcoming in *Synthese*.

[88] Ian Stewart, "A Budget of Trisections" (book review), *Mathematical Intelligencer*, Vol. 14, No. 1, 1992, pp. 73–7.

[89] John Tierney, "Behind Monty Hall's Doors: Puzzle, Debate and Answer?" *New York Times*, July 21, 1991, p. A1.

[90] John Tierney, "And Behind Door No. 1, a Fatal Flaw," *New York Times*, April 8, 2008, pp. C1.

[91] I. Todhunter, *A History of the Mathematical Theory of Probability from the Time of Pascal to That of Laplace*, Chelsea, New York, 1865.

[92] A. Tor, M. Bazerman, "Focusing Failures in Competitive Environments: Explaining Decision Errors in the Monty Hall Game, the Acquiring a Company Problem, and Multiparty Ultimatums," *Journal of Behavioral Decision Making*, Vol. 16, 2003, pp. 353–74.

[93] Elisabet Tubau, Diego Alonso, "Overcoming Illusory Inferences in a Probabilistic Counterintuitive problem: The Role of Explicit Representations," *Memory and Cognition*, Vol. 31, No. 4, 2003, pp. 596–607.

[94] John Venn, *The Logic of Chance*, Dover, Mineola, 2006.

[95] Marilyn vos Savant, *The Power of Logical Thinking*, St. Martin's Press, New York, 1996.

[96] S. Wechsler, L. G. Esteves, A. Simonis, C. Peixoto, "Indifference, Neutrality and Informativeness: Generalizing the Three Prisoners Paradox," *Synthese*, Vol. 143, No. 3, 2005, pp. 255–72.

[97] Ruth Weintraub, "A Paradox of Confirmation," *Erkenntnis*, Vol. 29, No. 2, Sept. 1988, pp. 169–80.

[98] Wikipedia entry, "The Monty Hall Problem," http://en.wikipedia.org/wiki/Monty Hall problem, last accessed January 20, 2009.

[99] Tung Yin, "The Probative Values and Pitfalls of Drug Courier Profiles as Probabilistic Evidence," *Texas Forum on Civil Liberties and Civil Rights*, Vol. 163, Summer/Fall 2000.

[100] Lawrence Zalcman, "The Prisoner's *Other* Dilemma," *Erkenntnis*, Vol. 34, No. 1, January 1991, pp. 83–85.

[101] Claudia Zander, Montserrat Casas, Angel Plastino, Angel R. Plastino, "Positive Operator Valued Measures and the Quantum Monty Hall Problem," *Anias da Academia Brasileira de Ciencias*, Vol. 78, No. 3, 2006, pp. 417–22.

Index